Alphonse Du Breuil, William Wardle

The Scientific and Profitable Culture of Fruit Trees

Alphonse Du Breuil, William Wardle

The Scientific and Profitable Culture of Fruit Trees

ISBN/EAN: 9783744678490

Printed in Europe, USA, Canada, Australia, Japan

Cover: Foto ©berggeist007 / pixelio.de

More available books at **www.hansebooks.com**

THE CULTURE OF FRUIT TREES.

THE SCIENTIFIC AND PROFITABLE CULTURE OF FRUIT TREES;

INCLUDING

CHOICE OF TREES, PLANTING, GRAFTING, TRAINING, RESTORATION OF UNFRUITFUL TREES, GATHERING AND PRESERVATION OF FRUIT, ETC.

FROM THE FRENCH OF
M. DU BREUIL,
PROFESSOR OF ARBORICULTURE IN THE CONSERVATOIRE IMPÉRIAL DES ARTS ET MÉTIERS.

ADAPTED FOR ENGLISH CULTIVATORS BY
WILLIAM WARDLE,
NURSERYMAN.

WITH 187 ENGRAVINGS.

Second Edition, carefully revised, with an Introduction by
GEORGE GLENNY,
AUTHOR OF "HANDBOOK TO THE FLOWER GARDEN," "GARDENING FOR THE MILLION," ETC., ETC.

LONDON:
LOCKWOOD & CO., 7, STATIONERS' HALL COURT.
LUDGATE HILL.
1872.

LONDON:
PRINTED BY VIRTUE AND CO.,
CITY ROAD.

INTRODUCTION

TO THE

SECOND EDITION.

The following treatise on the grafting, pruning, training, and general management of fruit trees having been submitted for my opinion, with a view to the issue of a new edition, with such alterations and improvements as I might suggest, I have carefully read it, and examined the practical directions throughout. There would have been no difficulty in making alterations and additions, since the subject is capable of enlargement to any extent; but the work is complete. It professes to direct the gardener in all the operations necessary, from the insertion of the graft to the completion of the tree, and the proper management through all its stages; and the instructions in all the various modes of grafting are full and comprehensive; many of them fanciful, and perhaps whimsical; but now that amateur gardening has become a fashionable amusement, the ingenious methods adopted by the French, who are in advance of us in fanciful grafting, budding, and training, are especially interesting. The instructions apply to all the useful fruits as well as the various plans of management; and the descriptive lists of all the most useful varieties are invaluable. The author has turned out a work that does him great credit, and the translator has shown that he perfectly understood the subject in all its bearings. The illustrations, nearly two hundred in number, are excellent.

GEORGE GLENNY.

Gipsy Hill, Norwood,
December, 1871.

TRANSLATOR'S PREFACE.

ONE of the ways of increasing our own knowledge is that of comparing it with the knowledge of others. It is not to be doubted that the various branches of Garden Practice and Literature have been well cultivated in England; and it is still less doubtful that these subjects have been as thoroughly studied in France, and expounded with unrivalled precision and clearness.

The present Work on the Cultivation of Fruit Trees is an evidence of this. It can scarcely be read by the most proficient Arboriculturist without imparting something to his previous knowledge, while to the many whose information is limited it cannot fail to be of great value.

It would be no small addition to our smaller pleasures to be able to grow for home use a good supply of our choicest fruits; and if a sufficient modicum of this knowledge could be imparted to our agricultural labourers and persons of limited means residing in the country, how much might be done to improve their incomings by growing good crops of the best descriptions of Pears, Peaches, &c., instead of a poor crop (if any at all) of comparatively worthless and unmarketable fruit. In France, this subject is considered of sufficient importance to be taught in schools and colleges, M. Du Breuil himself being a Professor of Arboriculture. The present Manual was honoured with the Prize of the Imperial Society of Horticulture.

The value of the Work to English Cultivators will be much enhanced by the additional notes of Mr. Wardle.

W.

CONTENTS.

	PAGE
GRAFTING	1
Grafting Mastic	3
Grafting by Approach, or Inarching	4
Branch Grafting	9
Cleft Grafting	12
Crown Grafting	15
Side Branch Grafting	16
Budding or Shield Grafting	19
ON PRUNING AND TRAINING	23
General Principles	24
Periods for Pruning	40
Instruments	42
Method of Pruning	43
THE PEAR	45
List of Choice Pears	47
Training and Pruning	52
Obtaining Fruit-branches	64
Maintaining ditto	77
Training in Vase or Goblet form	82
,, form of Column	86
,, as Double Contra Espaliers	89
,, Verrier's Palmette	95
Nailing up	103
Wire Trellis for Palmette	106
In Oblique Cordon	109
Trellis for ditto	116
Training in Vertical Cordon	119
Trellis for ditto	121
Standard Pear Trees	123
THE APPLE	127
Varieties	128
Training and Pruning	131
Low Apple Espalier	132
Mode of Inarching	133

	PAGE
THE PEACH	135
Varieties	136
Training in Palmette form	137
Trellis for ditto	145
Obtaining Fruit-branches	145
Training in Simple Oblique Cordon	168
Trellis for ditto	172
New mode of forming Fruit-branches	174
Advantages of ditto	183
THE PLUM	188
Varieties	189
Select list of	189
Training and Pruning	190
Training in Pyramidal form	191
Management of Fruit-branches	191
As Double Contra Espaliers	194
Verrier's Palmette	194
Single Oblique and Vertical Cordon	194
Standards	194
THE CHERRY	195
Varieties	195
Select list of	196
Pruning and Management	197
THE APRICOT	199
Varieties	199
Select list of	200
Pruning and Management	200
RESTORATION OF BADLY-TRAINED TREES	204
Standards	204
Espaliers	208
RENOVATION OF AGED TREES	209
Pyramidal form of	209
Espaliers	211
GENERAL DIRECTIONS FOR THE FRUIT GARDEN	213
Digging	213
Manuring	214
To counteract Excessive Dryness of Soil	215
Protection from the Late Frosts of Spring	216
GATHERING AND PRESERVATION OF FRUIT	221
Gathering	221
Preservation	222
Fruit-house	222
Care of Fruit	227

THE CULTURE

OF

FRUIT TREES.

GRAFTING.

THE PRINCIPAL METHODS OF GRAFTING TREES.—INSTRUMENTS, LIGATURES, AND MASTICS, OR GRAFTING CLAY.

NEARLY all fruit trees are propagated by grafting. It is necessary, therefore, to carefully study the practice of this operation; but we must limit our attention to those kinds of grafting that are really useful.

For the practice of grafting it is necessary to be provided with the following instruments. First a hand-saw (fig. 1), to be used in cutting the stems and branches that are too large to be cut with the knife. The saw-blade should be thin towards the back, A, and the teeth, B, wide, for cutting through green wood. A pruning-knife also is required for cutting the smaller stems and branches that are intended to receive the graft. This instrument will be described further on (p. 42).

A small wood mallet is also necessary to strike the back of the pruning-knife in making slits in thick stems for cleft grafting; also a small wooden wedge

2   FRUIT TREES.

to introduce into the opening while the graft is being placed in the slit. Lastly, a grafting or budding-knife is required (fig. 2). The spatula at the lower end of the knife should be made of hard wood, bone, or ivory. With this knife the lower part of the graft must be cut into proper shape, and shield grafting or budding performed.

Until the grafts become firmly grown to the wood, it

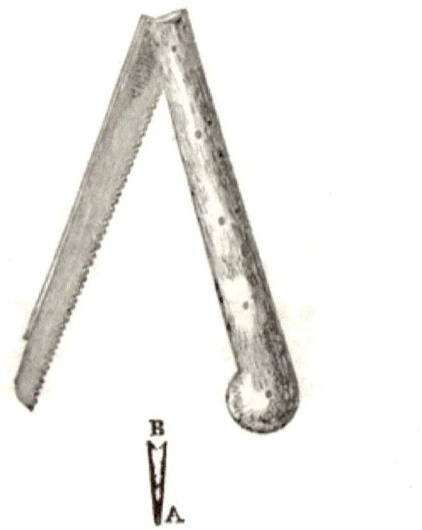
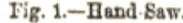

Fig. 1.—Hand-Saw.   Fig. 2.—Budding-Knife.

is of the utmost importance to keep them in their first position, and this is done by means of bandages composed of wool, drawn into coarse threads and slightly twisted, or strands of matting soaked in water, or of the bark of the willow or lime tree, made supple by soaking in water.

In order to protect the parts from the action of the atmosphere, various mastics are employed. One kind, called *Saint Fiacre's Ointment*, is chiefly composed of clay. Other descriptions, known as grafting mastics, are composed chiefly of resinous materials. The clay mastic has the inconvenience of being liable to be washed away by heavy showers, and of cracking when very dry, by which means it ceases to be a sufficient protection to the tender graft. When used upon apple trees, it also serves as a harbour for insects, which prove hurtful to the graft, and endanger the success of the operation.

Grafting mastics possess none of these inconveniences. They are used cold or hot. The mastic of which the following is the composition is always used hot.

For one hundred parts by weight:—

Black pitch . . .	28 parts.
Burgundy pitch . .	28 ,,
Beeswax	16 ,,
Grease	14 ,,
Yellow Ochre . .	14 ,,
	100

This composition must be used sufficiently warm to be liquid, but not so hot as to injure the tissues of the tree. A small brush is used to spread it with.

All the grafting mastics to be used cold were, until quite recently, in the form of a soft paste, and had the unpleasant inconvenience of sticking to the fingers of the operator. Hot mastics have, therefore, been preferred, notwithstanding the extra trouble of heating them. But now, M. L'homme Lefort, of Belleville, near Paris, has invented a liquid mastic that may be used cold. This mastic (the composition of which the inventor reserves the secret) is of the consistency of thin paste, which can be easily applied with a wooden spatula. In the course of a very few days it acquires an extraordinary degree of hardness, and is not affected by either sun or frost; humidity only hastens its solidification. This material, being sold at a moderate price, is likely to take the place of all other mastics.

The tree that is operated upon, and which receives the graft, is called the *subject* or *stock;* the small branch that is planted upon it is called the *graft* or *scion.*

The different methods of grafting, applicable to fruit trees, may be ranged in the three following groups:—

I.—GRAFTING BY APPROACH, OR INARCHING.

The peculiarity of this description of grafting is that the scion is not separated from the parent stem until after it has become completely united with the stock upon which it has been grafted; it is commonly performed in spring.

GRAFTING BY ORDINARY APPROACH.—This mode of grafting is employed to complete the number of lateral branches in a young tree, where there is no original insertion of either branch or bud, and where an incision would be useless. The part of the stem (fig. 3, A) which is deficient of one of its lateral branches, may be filled up by the aid of the branch B. An incision

GRAFTING. 5

must be made round the stem of the subject (A, fig. 4), immediately above the point where the branch B is to be grafted. This is done in order to retard the flow of sap from the roots; immediately below this, a vertical incision must be made, about 2 inches long, and of a width and depth equal to the diameter of the branch

Fig. 3.—Grafting by Approach, or Inarching.   Fig. 4.—Method of Cutting the Stock.   Fig. 5.—Cutting the Graft.

B, fig. 3. The branch must be cut at the point A, fig. 3, in a form (fig. 5) to fit exactly in the vertical incision, B (fig. 4), and the edges of the bark of the graft and of the subject must be brought into perfect contact. This being done, the parts must be brought together, and maintained in their positions by means of bandages, and covered with the grafting mastic.

By the following year, at the time of the winter pruning, the graft will have become firmly grown to its subject, and may then be cut off immediately below

Fig. 6.—English Method of Inarching or Grafting by Approach.

its point of attachment. The lower part, F, fig. 3, after trimming, will serve for one of the lateral branches as before.

English Method of Grafting by Approach, or Inarching (fig. 6).—Make, both upon the stock and graft, a longitudinal cut, in depth about one-third of the diameter; insert the one in the other; then place a clasp or bandage round the middle, to make the contact complete.

GRAFTING SHOOTS BY APPROACH (fig. 7).—In the preceding methods, the plants operated upon are at least a year old. In twig grafting, the graft, and sometimes even the stock, consist of tender herbaceous shoots. This method can only be practised from the middle of June until the beginning of August. The mode of operation differs also in other respects.

This kind of grafting is employed to greatest advantage upon peach trees and other stone fruits for filling up the vacant spaces on the fruit-branches, which constitute the lateral branches upon the principal branches.

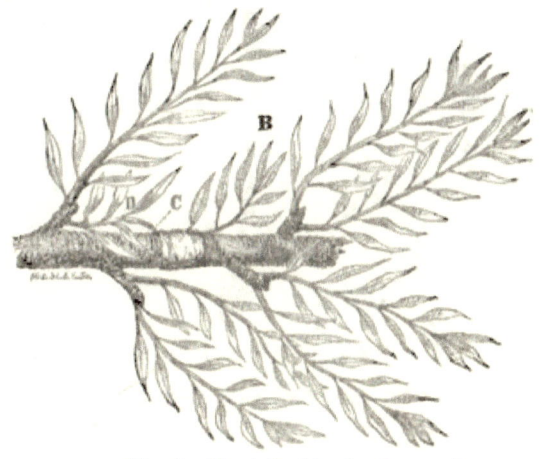

Fig. 7.—Shoot Grafting by Approach.

Suppose that there is a vacant space among the fruit-branches of a peach, C (fig. 7); the shoot B will serve to fill up the vacancy. For this purpose, make upon the branch, at the point C, an incision, about 1½ inch long, and terminated at each extremity by a transverse incision, C (fig. 8). The shoot B (fig. 7) must be cut as shown at D (fig. 8), and slipped under the raised bark, and the parts united by a ligature.

It is important that the graft should have a leaf at D, fig. 8, on the side opposite to the incision, which must be taken care of when applying the bandage.

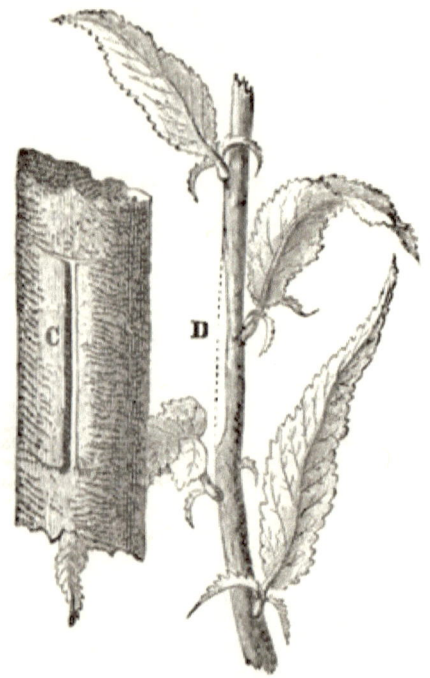

Fig. 8.—Shoot Grafting by Approach.

By the spring of the following year the union will be complete; nevertheless, it is desirable to delay the severance of the grafts from the lower part of the shoot until the second spring, otherwise many of them will be dried up. The proper time being arrived, the small branch which supplied the graft is cut at C (fig. 7), and the lower part of the branch D is allowed to grow as before.

If there are several continuous spaces upon the branch, and the smaller branch is sufficiently vigorous, there may be a succession of grafts upon each of the points A, A, A (fig. 9). When the grafting is com-

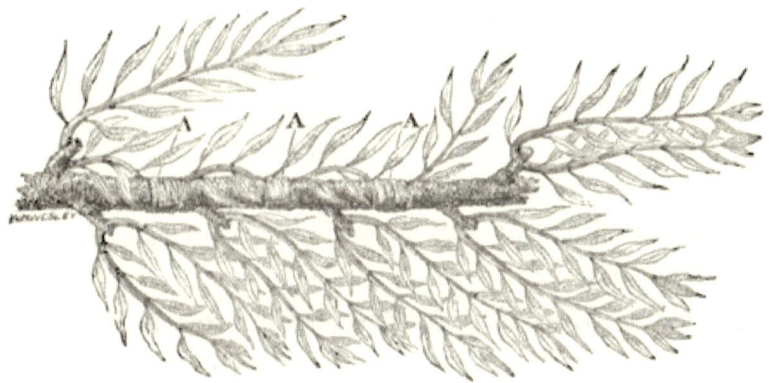

Fig. 9.—Multiplied Shoot Grafting.

plete, and the proper time arrived, the severance of each is effected immediately below each point of contact. It is better to allow eight or ten days to elapse between each grafting of the same shoot, to avoid injury to the future development of the grafts.

II.—BRANCH GRAFTING.

This mode of grafting is effected by means of branches, or parts of branches, previously separated from the parent stem.

To operate with success, it is essential:—1st, to select for grafts the most vigorous and perfectly ripened shoots of the *previous year*. That is, such only as have been developed sufficiently early in the preceding year to allow time for their wood to have become firm and well constituted before the first frosts of winter.

2nd. So arrange that the graft be in a less advanced stage of vegetation than the stock; otherwise, the graft, instead of finding a sufficient quantity of sap to develop its growth, will rapidly dry up. To prevent this, the grafts must be severed from their parent stem a month or two before grafting, and completely buried in the earth at the foot of a northern wall. They will retain their vitality in this situation, while their vegetation remains stationary, and that of the subject develops itself.

3rd. Perform the cutting operations very neatly, that the bark be not torn, or frayed at its edges.

4th. Place the graft upon the subject in such a manner as to ensure the perfect contact of the edges of the inner bark of the stock with the corresponding edges of the inner bark of the graft.

5th. Bind round the parts operated upon, and cover over with grafting mastic.

6th. Protect the grafts, during the first five days after grafting, from the action of the atmosphere and heat of the sun. A cone of white paper answers this purpose perfectly (fig. 10). The cap, however, is liable to harbour insects that eat the buds as soon as they open.

7th. Great care is requisite that the grafts, once placed, be never afterwards disturbed. The least shock is sufficient, when the graft has begun to grow to its stem, to destroy the success of the entire operation. The grafts most exposed to accidents are those placed upon the top of high trees, they being frequently broken by the larger birds perching upon them. To prevent accidents from this cause, a kind of perch, A (fig. 11), may be affixed, fastened to each side of the stem by two osier bandages. This possesses an additional advantage in the security it affords to the principal

Fig. 10.—Paper Cap for Grafts.   Fig. 11.—Graft Protector.

branches, B B, during the summer which develops the graft, the grafts being tied to the stock to prevent their being blown off by the wind.

8th. Lastly, it is necessary to be careful that the numerous branches which always grow from the head of the stock do not destroy the graft by absorbing all the sap. During the summer which follows grafting, the stem of the subject will be covered with young shoots. These must be taken off, but not until the graft has begun to grow, for until that period it has need of them to draw the sap towards itself. As soon as the graft begins to shoot, the buds growing from the lower part of the stem must be suppressed; afterwards, the buds higher up progressively; but those growing close about the graft must not be removed until the graft has attained a height of from 6 to 10 inches.

Branch grafting proper for fruit trees is of the three following kinds :—

1. CLEFT GRAFTING.—This requires a longitudinal incision in the wood of the stock. The operation is performed in spring, as soon as the buds of the stock begin to open.

Single Cleft Grafting (fig. 12).—The branch which serves for graft may be from 3 to 6 inches in length, according to the degree of vigour shown by the stock. Choose for the graft a branch with a bud upon its summit, A. Cut the lower part of the graft in a flat pointed form, to the length of $1\frac{1}{2}$ or 2 inches, commencing from the bud B. When the graft is thus prepared, cut the head of the stock horizontally across, and smooth it with a sharp knife; upon this cut make with a knife a vertical cleft, C, down the middle of the stem, about 2 inches long. The cleft must be kept open by a small wedge while the graft is being placed in it.

The top of the graft E, fig. 13, should be slightly inclined towards the centre of the stem, while the lower part, F, should project a little outwards, in order that the interior bark of the graft and of the subject may

GRAFTING. 13

be brought into exact contact at one point of their exposed surface. Finally, the graft must be bandaged, and the parts covered with a coating of grafting mastic. This method is employed for fruit trees either with high or low stems, provided the stems be not too thick.

Double Cleft Grafting (fig. 13).—This differs from the preceding by two grafts being placed in the cleft

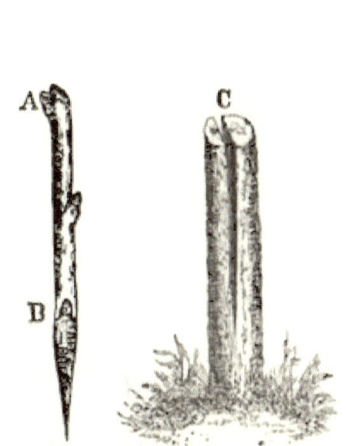

Fig. 12.—Single Cleft Graft.   Fig. 13.—Double Cleft Graft.

instead of one. This method is to be preferred when the thickness of the stock allows of it. The parts heal more quickly, and there is a better chance of success than with only one graft. Nevertheless, if both prove successful, we must not hesitate to suppress the least vigorous one as soon as the place is completely closed, especially in the case of tall trees; otherwise, the head of the tree being formed of two parts quite distinct from each other, the stem would be liable to be torn asunder when heavily laden with fruit, or exposed to the influence of high winds.

Bertemboise Cleft Grafting (fig. 14).—Cut the head of the stock on a slant, reserving at the highest point

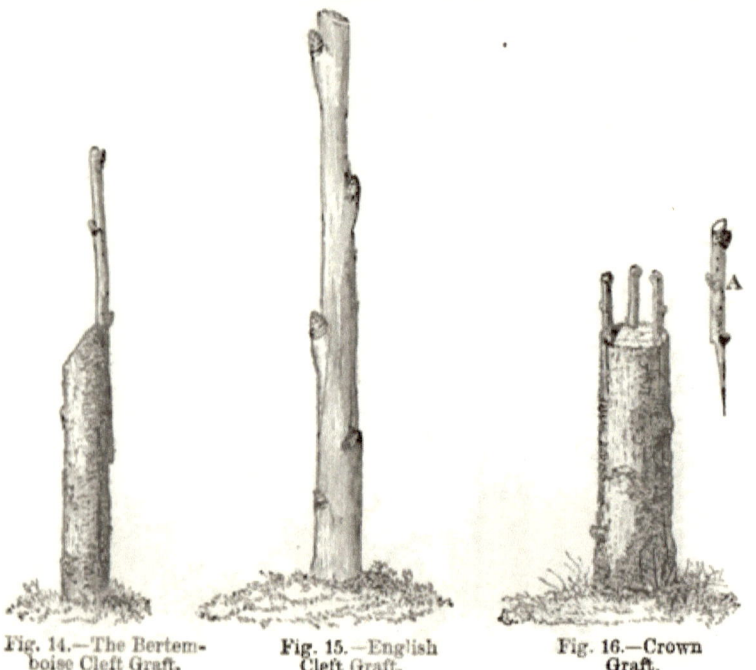

Fig. 14.—The Bertemboise Cleft Graft.   Fig. 15.—English Cleft Graft.   Fig. 16.—Crown Graft.

a small portion of flat surface, and operate as before. When the stock is not large enough to carry two grafts, this method is preferable to the two preceding ones. The point of union between the graft and stock will be better formed, and the oblique shape of the crown of the stock will dispose the sap to flow more freely towards the graft, which will develop more vigorously.

English Cleft Grafting (fig. 15).—Cut the stem of the stock on a long slant, and make a vertical cleft down it, commencing about a third of the distance from the top. Cut the lower part of the graft in a corresponding slant, and make a vertical cleft commencing about a third of the distance from the base; introduce the tongue of the graft into the cleft of the stock, in such a manner that the parts may be perfectly covered by each other, and so that the barks come into perfect contact on at least one side of the stem. This is a very ready and firm mode of grafting, and suitable for young stocks, the parts covering each other over their entire surface.

2. CROWN GRAFTING.—In this description of grafting the wood of the stock is not cleft, but the bark alone is cut vertically. The operation is performed when the buds are grown a third of an inch long.

Crown Grafting, Theophraste (fig. 16).—Cut horizontally the head of the stock, or the branches only of the second or third order, according to the age of the tree, at about 18 inches from their spring. Then cut through the bark to the wood in a vertical line about $2\frac{3}{4}$ inches long. Cut the lower part of the graft in a pointed form, A, with a notch on the upper part. Raise the bark of the stock, and introduce the graft between the bark and the wood in such a manner that the cut side of the graft fits close upon the wood. Surround with a bandage, and cover with the mastic.

Several grafts may be placed upon the same section of the wood, provided there be a clear space of about 3 inches between each of the grafts. This sort of graft is frequently used for aged trees, and when it is desirable to change the fruit for another variety.

Perfected Crown Grafting, Du Breuil (fig. 17).—The

head of the stock must be cut obliquely, then the bark cut vertically a little to the left of the top of the slant. The lower part of the graft must be cut in a pointed form, with a small tooth or notch at the thick end of the cut. A thin line of bark is then cut away upon the left side of the point end of the graft. The graft is inserted between the bark and the wood, so that the

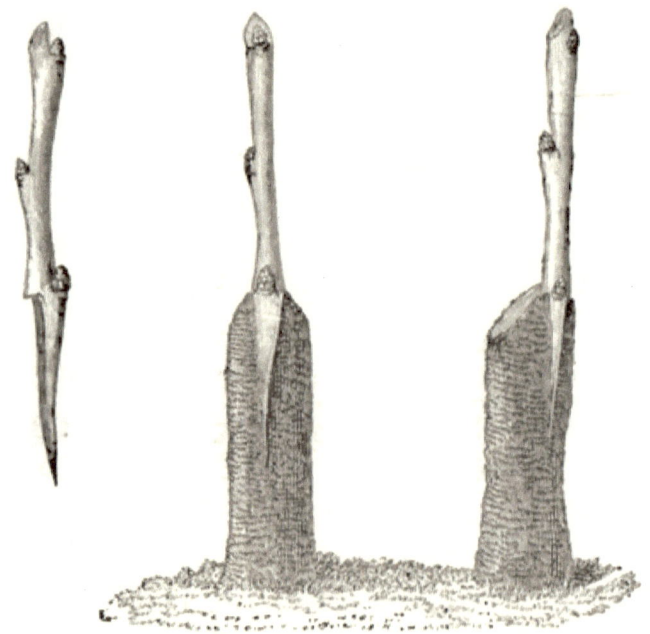

Fig. 17.—Du Breuil's Improved Crown Graft.

notch on the graft fits close down upon the top of the stock, and the point of the graft introduced under the bark of the right side only of the stock, fits closely against the unraised bark on the left side of the stock.

3. SIDE BRANCH GRAFTING.—In this kind of grafting it is not necessary to cut the head of the stock. The graft is attached to the side of the stem. It is

practised at the same period of the year as crown grafting. We shall explain only two varieties of this method.

The Richard Side Graft (fig. 18).—Choose for the graft a branch slightly arched, A; cut the lower part

Fig. 18.—The Richard Side Graft.

Fig. 19.—Lateral Fruit-Bud for Girardin Side Graft.

in a long slant. Make an incision, G, upon the bark of the stock in the form of a T. Make immediately above this another incision, B, penetrating to the exterior wood, for the purpose of arresting at this point the flow of sap from the roots. Raise the incised bark with the spatula of the grafting-knife, introduce the graft, bind round, and apply the mastic.

This mode of grafting is applied with advantage to the stems of *pépin** fruits, to restore the regularity of

---

* This word implies fruits that have "pips," and is used in that sense throughout the work.

their branches, when grafting by approach, or other methods, cannot be resorted to.

The Girardin Side Graft (figs. 19 to 22).—This kind of grafting, described by Professor Thouin, and popularised by M. Luiset, of Ecully, near Lyons, is practised thus:—Towards the end of August take from a tree, either of the same variety or a different one, some very small branches, each bearing a fruit-bud that would open in the following spring (fig. 19); in

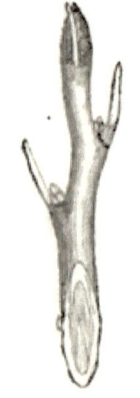  

Fig. 20.—Terminal Fruit-Bud for Side Graft. Fig. 21.—Cut upon the Branch for Reception of Side Graft. Fig. 22.—Girardin Side Graft.

selecting the grafts, give preference to terminal branches (fig. 20). Cut off the leaves, and cut the lower part to the form indicated in figs. 19 and 20; make upon the bark of the stem upon which the graft is to be inserted an incision, as shown on fig. 21, insert the graft underneath the bark, bind round, as shown on fig. 22, and cover the parts with mastic. The grafts become attached to the branch, expand their blossoms, and fructify in the following spring. This method may be practised at the beginning of April, but with less chance of success; it will then be necessary to detach the branches which bear the grafts, a month before, and to cover them over in a shady place till the time of grafting.

This mode of grafting is now frequently employed to replace fruit-branches that have disappeared from the large boughs, or to transform too vigorous but unfruitful branches into fruit-bearing ones.

III.—BUDDING OR SHIELD GRAFTING.

In grafts belonging to this group the piece cut from the bark to form the graft is most frequently in the form of a shield, A (fig. 23); this piece of bark must have upon it, near its centre, an eye or bud.

These grafts are especially used for young stocks, or

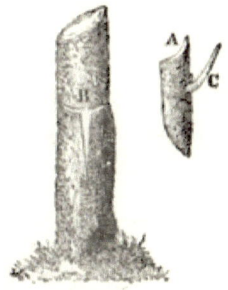
Fig. 23.—Shield Graft or Budding by Dormant Bud.

Fig. 24.—Inside of Cutting of Shield or Bud.

branches, of from one to four years of age, having thin, smooth, and tender bark. There are several varieties of shield grafting, but the two following will be found sufficient for general use.

SHIELD GRAFTING WITH DORMANT BUD (fig. 23).— This is practised from the end of July till the beginning of September, according to the state of the vegetation of the subject. The head of the stock must not be cut off until the following spring, when it will be seen whether the graft-bud has been successful. The following are the principal points to be observed in shield grafting:—

1st. Cut from the tree a branch having some leaves and eyes at its base, or buds well constituted; take off the leaves, leaving only a small piece of the stem, C (fig. 23), of one of them, to hold the shield A by between two fingers. Keep the grafts or buds, when thus prepared, in a dark, cold, and damp place, until the time that they are required for placing upon the stock.

2nd. Make an incision, B, upon the subject, in the form of a T, penetrating to the wood, and separate with the spatula the two lips of the bark towards the top.

3rd. Separate the shield from its branch in such a manner as to take off with the bark the smallest possible portion of the wood, preserving in every case the green tissue behind the bud, as shown at fig. 24. Unless this be attended to, the success of the graft is impossible.

4th. Slip in the shield between the bark and the wood by means of the incision B (fig. 23); then bring the edges of the bark together by means of a ligature, in such a manner that the base of the bud press closely to the wood of the stock; this is an essential point

5th. Some time after budding, look at the buds and slacken the bandages if they have become too tight.

6th. On the arrival of spring, if the buds have taken

effect, cut off the stem or branches of the subject about three inches above the bud; this is done to stimulate the development of the bud.

7th. When the buds begin to vegetate, support them by a stake (A, fig. 25), fastened against the stem of the

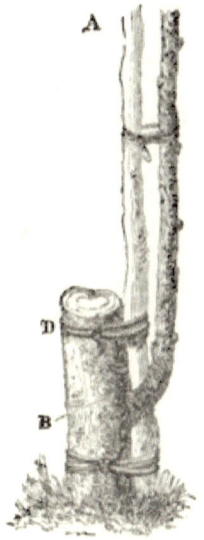
Fig. 25.—Support for Shield Graft.

Fig. 26.—Double Shield Graft.

subject, in order to protect them from the violence of the winds.

8th. Remove the shoots that grow from the stem of the subject, and follow the directions given for branch grafting.

9th. Cut the head of the subject D at the line B (fig. 25) in the winter following.

Shield grafts (or budding) are almost always used for young fruit trees. If they have not been successful (that no time may be lost), replace them by branch grafting in the following spring.

Double Shield Grafting or Budding (fig. 26).—Operate as before, placing upon the same stem or branch two or more shields. This will be found very useful in hastening the wood formations of young espaliers. To form a *palmette*,* the stock should receive three shield grafts, placed as shown in fig. 26. We thus gain a year for the formation of the wood.

* So called because the shape resembles the extended palm or hand, the fingers being spread out in the form of a fan.

ON PRUNING AND TRAINING.

UTILITY OF TRAINING.

Training, properly applied to fruit trees, gives the following result:—

1. It enables us to impart to trees a form suitable to the place they are intended to occupy. Thus we may desire to give to standards (trees that are not nailed against a wall or trellis) the pyramidal form, or the form of a vase.

Trees trained in these forms give larger and more abundant fruit than when left to themselves; and when become tall trees, they occupy less space than others. Training applied to espaliers makes the trees develop their wood in a regular and systematic manner, and compels them *to occupy usefully the whole surface of the wall* or space assigned to them.

2. By means of training, each of the principal branches of the tree is furnished with fruit-branches throughout its full extent. This result is most remarkable in the stone fruits, and especially in the peach, the branches of which, if not trained, would rapidly degenerate, and grow only towards the top.

3. Training renders the fructification more equal;

for in removing every year the superabundant buds and branches, we contribute to the formation of new fruit-buds for the next year; preventing the sap from being wasted upon the parts cut away.

4. Training conduces to the production of larger fruit and of finer quality. In fact, the greater part of the liquid nourishment which would have fed the suppressed parts is turned to the advantage of the fruits that are retained.

GENERAL PRINCIPLES OF TRAINING.

The wood of trees ought to be perfectly symmetrical. This regularity has not only for its purpose to render the trees pleasant objects to look at, but, most of all, to make them occupy with regularity, and without loss of space, the walls or borders where they are planted. It also promotes equality of vegetation throughout the tree, by preventing the sap from being drawn more to one side than another.

The permanency of form of trained trees is dependent upon the equal diffusion of sap being maintained throughout the whole extent of their branches.

In fruit trees left to themselves, and entirely untrained, the sap distributes itself equally, because the trees take the forms most in harmony with the natural tendency of the sap. In trees, however, that are submitted to training, the forms imposed upon them necessitate the development of branches at the base of the stem more or less large and numerous. The sap naturally tends towards the summit of the stem; it follows, therefore, that the lower branches become weak, and soon die off, and the form that had been obtained disappears, to be replaced by the natural disposition of the tree; that is, a naked stem carrying a head more or less voluminous. It is, therefore, indispensable to change the natural current of the sap, and thus maintain the direction towards each of the points to which it is desired to train the branches.

We suppose an espalier (fig. 27) in which the

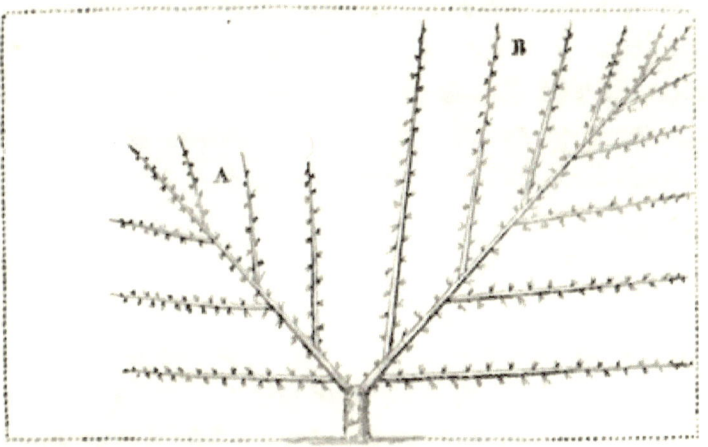

Fig. 27.—Espalier, in which the Circulation of the Sap is irregular.

equilibrium of vegetation has been broken; in order to retard the vegetation in the direction in which the sap flows too abundantly, and to favour those parts where its flow is insufficient, we employ the following means:—

Prune the strong branches, B, short; but allow the weaker ones, A, to grow long. We know that the sap is drawn by the leaves; therefore, in suppressing upon the most vigorous parts the greater number of wood-buds, we deprive those parts of the leaves that the buds would have developed, and in consequence of doing so, the sap flows there in less abundance, and the vegetation is diminished. By allowing, on the contrary, a considerable number of wood-producing buds to remain upon the weak parts, it will become covered with a large number of leaves, and an abundant vegetation.

Depress the strong parts of the tree, and elevate the weak branches. The sap on its way to the roots acts with greater force upon the branch extensions, in proportion to their being in a vertical direction; the branches will therefore push out with more force in the weak parts that have been raised nearer to a vertical line, and the numerous leaves that they will develop will draw the sap in greater quantity from the strong parts that have been inclined towards a horizontal direction.

Suppress the useless buds upon the strong parts as early as possible, and practise this suppression as late as possible upon the weak parts. The fewer shoots upon the branch, the fewer will be the leaves, and of course the smaller the quantity of sap that will be drawn there. By allowing the shoots to remain upon the weak part as long as possible, the sap will be drawn there in greater abundance, and when they come to be cut off, the sap, having set in to that side, will still continue to flow there. This is only applicable to espaliers, and particularly to peaches, from which it is often necessary to take a number of the buds.

Suppress very early the herbaceous extremities of the strong part, but practise the suppression as late as possible upon the weak part; taking off only the most vigorous shoots, and those that must in any case be removed on account of the positions they occupy. This arrests the vegetation of the strong part; it is applicable both to standards and espaliers.

Nail up very early, and very close to the wall or trellis, the strong part, but delay doing so to the weak part; we thus retard the circulation towards the first, and promote it towards the second. This is only practicable for espaliers and wall trees.

Suppress a number of the leaves upon the strong side. By removing a portion of the leaves on that side of the tree, we impede the sap, and prevent its too abundant flow in that direction. It will be necessary to suppress only a number of leaves proportionate to the difference in vigour of the two sides of the tree, and to choose those upon the most vigorous branches. The leaves must not be torn off, but cut, leaving the petiole, or leaf-stem, upon the branch.

Allow as large a quantity of the fruit as possible to remain upon the strong side, and suppress all upon the weak side. We have already explained that the fruit has the property of drawing to itself the sap from the roots, and of absorbing it entirely in its growth; by this means, all the sap drawn to the strong side will be absorbed by the fruits, and that part will be less developed than the weak side.

Soften all the green parts on the weak side with a solution of sulphate of iron. This solution, in a proportion of twenty-four grains to a pint of water, applied after sunset, is absorbed by the leaves, and powerfully stimulates their action in drawing the sap from the roots.

Bring forward the weak side from the wall, and keep the strong side close to it. By bringing the weak part forward, away from the wall, we allow the branches to receive the light on both sides. As light is the element which determines the functions of the leaves, and their action upon the sap, this part will vegetate with more vigour than the part which is only exposed to the action of the light upon one of its sides. This is applicable only to wall trees; and is only to be resorted to towards the month of May, by which time there is nothing to fear from the storms of winter, and the tree is able to support itself apart from the wall.

Place a covering upon the strong part, so as to deprive it of the light. By this method we obtain the same result, but in a more complete manner. This is not to be resorted to unless the preceding method has proved insufficient, for if it should happen that the part be kept too long from the light, it will soon lose all its leaves. To avoid this, we must not keep it covered for more than eight or twelve days, and must then take advantage of a cloudy day to remove the covering.

The different expedients we have described may be employed in succession, and in the order set down, until the desired result has been obtained.

The sap develops the branches much more vigorously upon a branch cut short, than upon one left long.

It is evident that if the sap acts only upon one or two branches, it makes them develop with much greater vigour than if it is divided among fifteen or twenty. If, therefore, we desire to obtain more wood, we must prune the branches down a great deal, because vigorous shoots develop very few flower-buds; if, on the contrary, we wish to develop fruit-branches, we must be careful to cut them down very little, because the least vigorous branches are the most charged with fruit blossoms. Another application of this principle is in the case of a tree that has been exhausted by bearing a too great quantity of fruit; we re-establish its vigour by pruning it short for a year.

This last application may appear to contradict what has been advanced in a former paragraph (p. 25), but the contradiction is only in appearance. In the first instance, we only cut short certain branches of the tree; and so far diminish, to the profit of those left to grow to a greater length, the powers of absorption that they exercise upon the sap from the roots. The shoots which they develop are certainly more vigorous than those growing upon the long branches, but they are less so than if all the branches of the tree were subjected to the same suppression, for one part of the sap due to them is turned to the profit of the numerous shoots growing upon the long branches, and the vigour of which is thus augmented. In a word, the shoots upon the long branches are not so vigorous as those upon the branches cut short, but they are much more numerous, and determine the formation of a great mass of woody tissue and buds, which do not fail to weaken the strong part to the profit of the feeble one.

But when the re-establishment of an exhausted tree is in question, the circumstances are altogether different. Instead of pruning down certain branches only, we submit all the branches upon the tree to the same treatment, and the sap, not being drawn in too great abundance either to one side or the other, acts with equal intensity, and promotes the vigorous development of each separate branch: all this tends to the formation of fresh wood and bark, more ample and well constituted than the first, so that the new root extensions better fulfil their functions, the tree recovers its former vigour, and in due time recommences fruit-bearing.

The preceding remarks explain the cause of the different results obtained by this operation, according to the manner in which it is practised, and ought to put an end to the differences which exist upon the subject among different cultivators.

The sap has always a tendency to flow towards the extremity of the branches, and to make the terminal bud develop with more vigour than the lateral ones.

According to this principle, whenever we wish to give greater length to a branch, we must train upon one vigorous terminal bud, and not allow any other shoot to grow that would draw away the action of the sap.

The more the sap is retarded in its circulation, the less wood, and the more fruit-buds will it develop.

Trees do not begin to form their flower-buds until they have acquired a certain development. Before these flower-buds appear, it is necessary that the sap circulate slowly, and that it by this means undergoes a more complete elaboration in the leaves; without this it can only grow wood-producing buds. When trees have acquired a certain degree of development, the rapidity of the circulation is checked by the extent and broken character of the ramifications through which the sap has to run; it is only then that the flower-buds begin to form. The appearance of these organs is so much due to the diminished action of the sap upon the branches, that trees never have more flower-buds than when suffering in this way.

The following operations, in the order set down, will tend to retard the action of the sap, and cause a greater quantity of fruit-bearing spurs upon the trees. Allow the branch to extend itself by training the wood very long, and the result will be a less vigorous growth which is more conducive to fruit-bearing.

Apply to the branches which grow from the successive extensions of the wood, and also to those which spring from them, the operations calculated to diminish their vigour. These operations are, for shoots, pinching and twisting, and for the branches, breaking, either complete or partial. These mutilations, which we shall describe further on, have for their object to diminish the vigour both of these smaller and larger branches, by forcing the sap to concentrate its action upon the development of new wood.

Practise the winter pruning very late in spring, when the shoots have obtained a length of 1½ inch. It results from this late pruning, that a greater portion of the sap is dispersed to the advantage of the other parts of the branches. These being at this moment checked, the shoots at the base push less vigorously than if that loss of sap had not taken place, and are more easily put into fruit-bearing. This mode of operating, and also the following, should not be resorted to, except for trees of such vigour that the preceding methods have proved insufficient to put them in fruit-bearing.

Apply to the larger branches a certain number of Girardin's side grafts (figs. 19 to 22). These fruit-spur grafts when fruiting will absorb a considerable part of the superabundant sap. We shall see from this time a considerable number of flower-buds form themselves upon the trees. This method is only to be employed for *pépin* fruits.

Arch all the branches, so that a part of their extent be directed towards the sun. The sap acts with more force upon the development of shoots if they are attached to branches growing in the nearest approach to a vertical line; it follows, therefore, that arching the branches tends to diminish their vigour, and promotes fruit-bearing. When this result has been obtained, it will be advisable to replace these branches in their former position, lest the tree become exhausted by a superabundant production of fruit. Fig. 28 is a pyramidal tree with the branches arched.

PRUNING AND TRAINING.

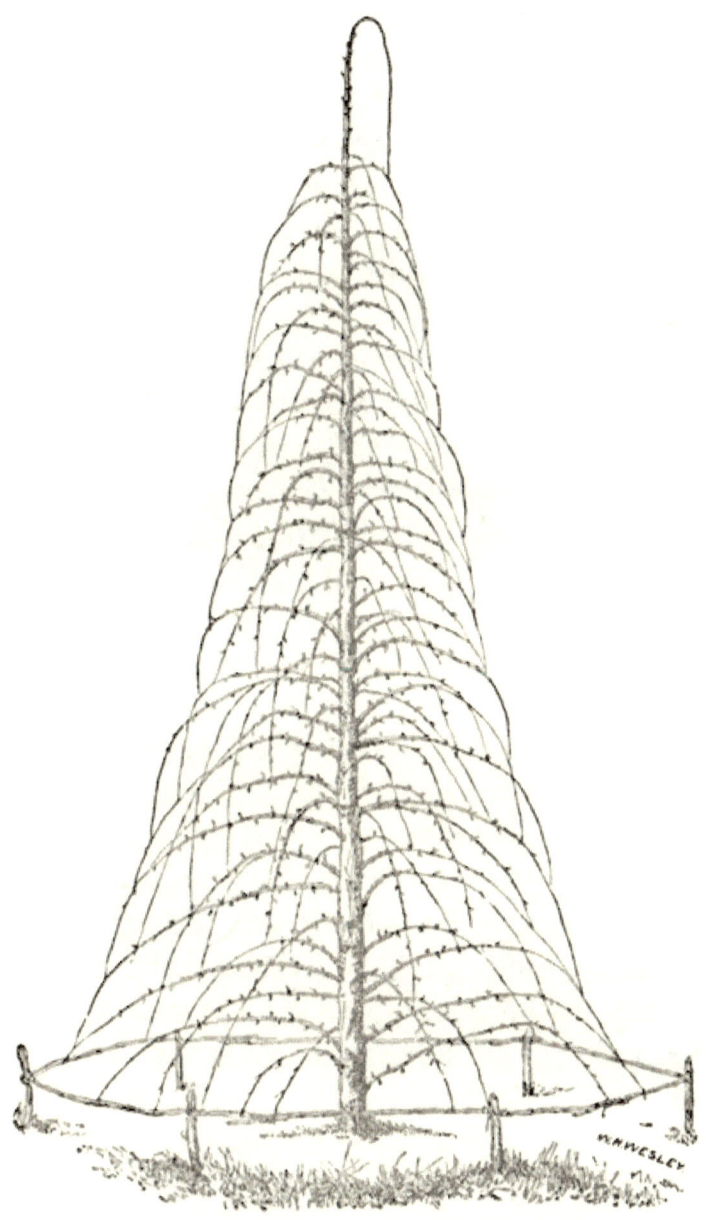

Fig. 23.—Pyramidal Pear Tree, with Arched Branches.

In the month of February, make an annular incision with the hand-saw, near the base of the stem, rather less than ¼ of an inch in width, and sufficiently deep to penetrate the exterior layer of the wood. The sap ascends from the roots to the leaves, passing through the sap vessels embedded in the exterior layer of wood. The incision has the effect of retarding the ascension of the sap; the branches acquire less vigour; and the tree forms fruit.

At the spring of the year, uncover the foot of the tree in such a manner as to expose the principal roots throughout nearly their entire extent, and allow them to remain in this state during the summer. The exposure of the roots to the action of the air and light retards their fruition, and thus diminishes the vigour of the tree, and determines its fruit-bearing.

Uncover the foot of the tree at spring, and cut away part of the roots, and then replace the earth. This operation, more energetic than the preceding, produces the same results, but must be resorted to with caution, to avoid injury to the tree.

Transplant the trees at the end of autumn, but with great care so as to preserve all the roots. This affords the same results as the preceding. The displacement of the tree has the effect of weakening it; in the following year it will put forth a great number of fruit-buds.

Every method which conduces to diminish the vigour of the wood, and to make the sap flow to the fruit, tends to augment the size of the fruit.

The fruit and the branches have the property of drawing to themselves the sap from the roots. If, therefore, the shoots are numerous and strong, it follows that they absorb nearly all the sap, to the injury of the fruit, which remains small. This explains how it occurs that, other things being equal, fruits are smaller upon vigorous trees than upon weaker ones; and we also understand from this how it is that, the growth of the fruit being determined by the sap, the fruit becomes much larger if the sap flows to it freely.

The following operations have for their object to increase the size of the fruit:—

Graft upon stocks of a less rigorous species than the scions. If the stocks are too vigorous, the shoots will absorb nearly the whole of the sap, to the injury of the fruit. Pears grafted upon quinces, apples upon paradise stocks, produce, other things being equal, larger fruit than when grafted upon pear stocks.

Apply to the trees a suitable winter pruning; that is, do not leave upon them more branches, or parts of branches, than are requisite for the symmetrical development of the tree, and the formation of fruit-bearing branches. This tends to concentrate the sap upon the parts retained, and consequently upon the fruit. Trees left to their natural growth always produce smaller fruit than those submitted to suitable pruning.

Make fruit-spurs to grow close upon the branches, by pruning them as short as possible. By this means the fruit will be attached very close to the wood, will receive the direct influence of the sap, and acquire a large development.

Cut the branches very close when the flower-buds are formed. This concentrates the sap upon a smaller extent of wood, and the fruit receives, in consequence, a larger supply.

Mutilate the summer shoots by repeatedly pinching off those shoots that are not required for the development of the size of the tree. This mutilation, which is performed by repeated pinchings, prevents the shoots absorbing a too large supply of sap, which then remains, to the advantage of the fruit.

When the fruits have attained a fifth degree of their development, suppress a further number of them. The fruits left upon the tree absorb the sap of those taken away, and therefore become much larger. There will be a smaller number, but the same weight, which is always to be preferred.

Make an annular incision upon the fruit-bearing branches at the time they expand their blossoms; the incision must not be wider than $\frac{1}{16}$ of an inch. Experience continually demonstrates, that following such incision, the fruit becomes much larger, and ripens better. Many attempts have been made to explain the cause, but none are satisfactory; the fact, however, is certain. Stone fruits, and vines especially, are the better for this operation being applied to them.

Graft some of the fruit-branches of vigorous trees with the Girardin side graft. This kind of graft produces an effect similar to the annular incision. The fruit is always larger than upon the other branches; the cause is doubtless the same.

Place under the fruits, during their growth, a support,

to prevent their stretching or twisting their footstalks. The sap reaches the fruit through the vessels which traverse the stalk. If left without support it will often happen that the fruit grows unequally, and a twisting movement of the stalk follows, which injures the sap vessels. Besides, the weight of the fruit alone, hanging upon its stem, stretches the sap vessels, and diminishes their diameter. When the fruits are supported, the sap penetrates more freely, and their size is augmented accordingly.

Keep the fruits in their normal position during the entire period of their development, that is, with the fruit-stem lowermost. The sap acts with greater force when it flows upwards; a vertical position, therefore, of the stalk causes the sap to ascend more easily and in greater quantity, and the fruit will become larger.

Place the fruits under the shade of the leaves during the entire period of their growth. The action of strong light and heat has the effect of hardening the tissues, and destroying their elasticity, and consequently the power of extension in yielding to the action of the sap. If a young fruit be exposed to the power of the sun, it will be smaller than one shaded by the leaves, because its skin will be hardened, and not give way to the tendency of the sap to expand it. Fruits, when arrived at their full size, will be greatly improved by exposure to the sun, as it will impart colour and a finer flavour, and ripen them in greater perfection.

Apply to the young fruits a solution of sulphate of iron. We have already seen that a solution of sulphate of iron applied to the leaves stimulates their powers of absorbing sap. The thought occurred to apply the solution to the fruits, and the effect in increasing their size was extraordinary. The solution should be in the proportion of 24 grains to a pint of water. Apply it only when the fruit is cool; repeat the operation three times; viz., when the fruits have obtained a fourth part of their development, when they are a little larger, and again when they are three parts grown. This solution excites their powers of absorption, and they draw to themselves a large quantity of sap which would otherwise flow to the leaves, and they then become larger fruit.

Graft by approach a small shoot upon the peduncle or fruit-spur, to which the fruit is attached when it has attained a third part of its development. It has been found that consequent on this operation the fruit becomes larger, doubtless because the graft draws to the peduncle a larger quantity of sap.

The leaves serve the important purpose of elaborating the sap of the roots and preparing it for the proper nourishment of the tree, and the formation of buds upon the boughs. A tree therefore that is deprived of its leaves is in danger of perishing.

It is therefore essential to guard against removing too many of the leaves under pretext of placing the fruit more immediately under the action of the sun, for the trees, deprived of a part of their organs of nutrition, will cease both growing and fruit-bearing. Besides, branches stripped of their leaves produce ill-formed buds, which are only succeeded the following year by a weak and languishing vegetation.

When the ramifications of the tree have arrived at their second year, the buds that are still undeveloped will remain so, except under the influence of very close pruning. Peaches seldom yield to this operation.

It is necessary, therefore, to prune so as to determine the development of these dormant buds upon the branch extensions, and to take care of the shoots which result therefrom. Without this precaution the middle part of the tree might remain bare and unproductive, and there would soon be no remedy, for it is impossible to develop buds that have remained long dormant. We obtain the development of all such buds by cutting away each year a certain portion of the new extensions of the wood.

The yearly extensions of wood should be shortened more or less, as the branches approach a vertical line, or the contrary.

The sap acting upwards from the base of the tree, if a branch grows in a vertical direction, the buds remain dormant upon two-thirds of its length from the base. To prevent this, it is necessary to suppress at least one-half the length of this branch. If it is inclined to an angle of 45 degrees, the sap acts with less force upon the buds at the upper part of the branch, but it will develop much too great a number at the same time; only the lower third of the branch will remain unfurnished with buds. It will be sufficient in this case to suppress only a third part of the branch.

Lastly, if the branch grows horizontally, it may be left entire, for in this case the sap will act equally, and buds will spring from every part of the bough, from one end to the other.

The most suitable Periods for Pruning.

The various operations of pruning fruit trees are practised at two different periods of the year. Those comprehended under the name of *winter pruning* are performed while vegetation is at rest; the others, called summer pruning, at various periods of vegetation.

We shall first point out the most favourable period for performing the winter pruning.

Winter Pruning.—The best time is that which follows the severe frosts, and which precedes the first movement of vegetation, towards the month of February.

If the pruning be performed before the severe frosts, the cut portions of the bough will be exposed to the action of the air, damp, and frosts, long before the first movement of the sap, which is necessary to cicatrise the parts, and it frequently follows that the terminal bud reserved at the top of the branch is by this means destroyed.

These accidents occur also if the pruning be attempted during severe frosts. The instruments cut the frozen wood with difficulty, the parts are torn and do not heal, the injury descends below the nearest bud, which is destroyed.

If we wait until the buds begin to open, the consequences are still more serious. The sap is spread throughout every part of the tree, and that which was absorbed by the suppressed branches is lost. Besides, in pruning so late, there is great danger of breaking off a number of the buds. Finally, the sap, in rushing back from the summit to the base, will burst the sap vessels, and cause canker or gum.

The winter pruning, in February, is always important for peaches; the buds on the lower parts of the fruit-branches frequently remaining dormant for want of a sufficiently powerful action of the sap.

By early pruning, the sap is made to act upon the buds unfavourably situated upon the tree, brings them out, and also develops latent buds upon the old wood.

We are thus, by early pruning, enabled to prevent the middle part of the tree from becoming bare and unproductive.

By late pruning, waiting even to the period when the shoots begin to lengthen, we may operate with advantage upon trees that possess too much vigour, and which would not otherwise be easily put into a fruit-bearing condition. One part of the action of the sap is thus dispensed to the parts cut away, and acts with less force towards the reserved buds, which thus more readily assume the character of fruit-bearing ones.

In southern parts of the country [France], however, where vegetation is early, the pruning must necessarily be performed before winter.

If, also, we have such a number of trees that we shall not be able to complete the pruning of the whole in February, rather than go beyond that time it will be better to anticipate it, by pruning the fruit-branches before winter, and leaving the rest until February.

It will in every case be necessary, in pruning, to follow the order of vegetation of the different species, pruning first the apricots, next the peaches, plums, cherries, pears, apples, and lastly the vines.

Summer Pruning. — The operations of summer pruning are practised while vegetation is entire; but the precise moment can only be determined by the actual state of vegetation of the parts of the tree that require pruning. In order to describe these indications with greater clearness, and to avoid useless repetition, we shall defer our remarks on this head until we come to apply these operations to the various species of fruit trees of which we have to speak.

Instruments required for Pruning.

The pruning-knife (fig. 29) is the oldest and best of the instruments employed in pruning. The blade should be sufficiently curved, but not so much as to form a right angle, for in that case it would be as difficult to cut with as if the blade were nearly straight. The haft should be large enough to fill the hand. It is requisite to be provided with two knives, a very strong one for winter pruning, and the other much smaller for summer operations.

It has been proposed to substitute the sector (fig. 30) or pruning shears for the pruning-knife. The sector offers the advantage of effecting its work more promptly than the knife, but it flattens the wood at the point of section, detaches the bark for a short distance below the place, and the end of the branch withers instead of healing; the injury often extends below the terminal bud, which is thus destroyed. To obviate this, it is necessary to cut about half-an-inch above the bud;

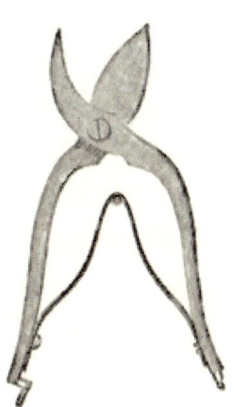

Fig. 29.—Pruning-Knife.   Fig. 30.—Sector.

this leaves a short piece of dried branch which it is necessary to cut off the following year, which unnecessarily multiplies the operations. The knife is, in our opinion, the better instrument. If, however, we sometimes use the sector, we should place it in such a manner upon the branch as to cut away nearly all the part injured by the pressure.

The operator should also provide himself with a small hand-saw (fig. 1, page 2).

Method of Pruning.

The manner of cutting the branches is far from being unimportant. If we wish to shorten a branch (fig. 31), we make the amputation as near as possible to a bud, but not so near as to injure it. We place the pruning-knife exactly opposite the bud, and cut in a slanting direction, in the line A B, coming out a little above the bud. By this means the bud remains uninjured, and the part more readily buds.

If we cut higher than this point, in the line A B (fig. 32), the wood dies down to the line C, leaving a little dried stump, which has to be cut off the following year. If we follow the line A B (fig. 33), the bud is weakened, and its development will be much less vigorous.

When it is desired to cut away a branch entirely, the cut is made quite at the base, always leaving, how-

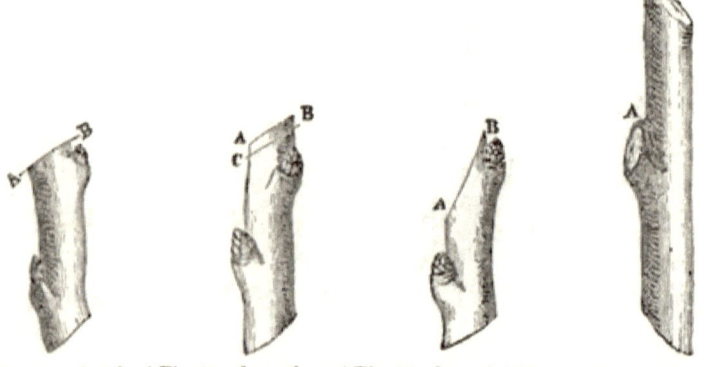

Fig. 31.—Cut for branch in pruning | Fig. 32.—branch cut too far from the bud | Fig. 33.—branch cut too slanting | Fig. 34—Complete suppression of branch.

ever, a very small stump, A (fig. 34). We thus make a smaller wound, and it heals more rapidly than if it had been cut closer to the stem.

If a branch is too large to be cut with the knife, and the saw has to be applied, it will be necessary to plane or smooth off the rough part left by the saw, otherwise the place will heal badly. If the parts that have been cut are large, it will be desirable to cover with the grafting mastic.

THE PEAR.

Soil.—The pear takes a deep clayey flinty soil, rather cool, but not humid. In every case where this is not the native character of the soil, it must be made so by the mixture of other materials, and by digging to at least a yard in depth. If the under soil is too damp, it must be properly drained.

Choice of Trees.—If the pears be taken from the nursery ready grafted, choose healthy vigorous trees of one, or at least of two years' grafting. More aged ones than these take root less freely, and their vegetation is always retarded. The stocks might be planted in the nursery for grafting them the year following, and planting them out where they are to remain the next year.

Grafting.—Pears are most frequently grafted upon pear stocks obtained by sowing the *pépins;* they are also grafted upon quince stocks. The first produces the most vigorous and durable trees, but the quince stocks more rapidly come to fruit-bearing.

We prefer pear stocks for dry and rather poor soils, and quince for ground of richer and better quality. The varieties that have a less vigorous habit should in every case be planted upon pear stocks. We point out which these are in the list on page 47.

The grafts to be employed are, the shield graft or budding, the English cleft graft, and the protected crown graft. The shield graft or budding must be practised in August, on young wood of the current year's growth; the crown and cleft grafts are used for more advanced trees, or to replace a shield graft that does not grow. The crown graft mutilates the tree less than cleft grafting.

Varieties.—We are acquainted at the present time with more than five hundred varieties of pears. But the whole of them are far from being equally valuable. We shall here point out only some of the best for each month of the year, and place opposite to each name certain directions necessary to their culture. The names in italics indicate the principal synonymes.

DU BREUIL'S LIST OF CHOICE PEARS.

NAME OF THE VARIETY AND SYNONYMES.	DESCRIPTION, OR ENGLISH NAME.	TIME WHEN FRUIT RIPENS.	TRAINING.	ASPECT.	OBSERVATIONS.
Doyenné de juillet...... *Roi Jolimont*	*Summer Doyenné, or Doyenné d'Eté* ...... Same as above	End of July	Standard		An excellent early pear. Good bearer on any aspect
Beurré Giffart...... Epargne ...... *Belle Verge* *Poire de Seigneur* *Cuillette* *Poire de la table des princes* *Saint-Sanson* *Beurré de Paris* *Grosse cuisse Madame d'Eté* *Roland* *Chopine*	...... *Jargonelle* ...... A fine old pear, does well on the quince stock in the North of England, and is known by all these names as synonymes, and will do on any aspect on a wall, but does not ripen well as a standard	August August	Standard Stand. Espa.	E W	Graft upon a pear stock. Dry ground; assumes the Pyramidal form with difficulty
Beurré Beaumont ...... *Rey Waci* *Bermont*	*Besi de St. Wai* ...... Does not ripen well, except on quince stock, in the North of England	December	Standard		Good bearer on quince; worthless as a standard on pear stock
Beurré d'Amanlis ...... *Wilhelmine* *Poire Hubard*	One of our best autumn pears, ripens in September, and keeps for a month after	September	Standard		Great bearer on quince
Bon-chrétien (Williams's)...... *De Lavault*	......	Aug. & Sept.	Standard		Graft upon pear stock for wall, and quince for open ground
Beau présent d'Artois...... *Présent royal de Naples*	*Beau présent d'Artois* ...... Fruit large and juicy, moderate flavour	September	Standard		
Jalousie de Fontenay-Vendée *Belle d'Esparmes*	*Jalousie de Fontenay* ...... Ripens in October and November; requires a south wall; does well as dwarf bush on quince	Sept. & Nov.	Standard		Graft upon pear stock for wall, and on quince for dwarfs

NAME OF THE VARIETY AND SYNONYMS.	DESCRIPTION, OR ENGLISH NAME.	TIME WHEN FRUIT RIPENS.	TRAINING.	ASPECT.		OBSERVATIONS.
Seigneur d'Esperen *Bergamote fiérée* *Bergamote lucrative*	*Fondante d'Automne,* or *Beurré d'Albert* Fine large handsome fruit, with a delicious flavour; does best on quince as dwarf standard	Sept. & Oct.	Stand.			Graft upon pear stock, or on quince for dwarf bushes
Professeur Du Breuil		September	Stand. Espa.	E		
Beurré d'Angleterre *Bec d'oiseau*	*Brown Beurré of the English* An excellent pear	October	Standard	W	S	Graft upon the small sugar pear, itself grafted on a quince
Doyenné doré *Doyenné blanc* *Saint-Michel* *Poire de neige*	*White Doyenné,* or *White Autumn Beurré* Tree is hardy, and fruits well, either standard or on quince	Sept. & Oct.	Standard			Does well on quince for dwarfs
Louise Bonne d'Avy *Louise Bonne de Jersey*	The finest autumn pear in existence; does well as a Pyramid on the quince	October	Stand. Espa.	E		Dry ground; fruits well in almost any situation
Beurré gris *Poire d'Amboise* *Beurré roux* *Beurré doré*	*Brown Beurré* A well-known pear, of great merit	October	Espalier	E	W	Graft upon pear stock
Beurré d'Avy *Beurré Spence* *Fondante des bois* *Belle de Flandre*	*Flemish Beauty* A very fine-flavoured pear, and generally ripens well	Sept. & Oct.	Stand. Espa.	E	W	Does well as a standard from a wall; takes well on the quince; a great bearer; very suitable for a market gardener
Beurré des Charneuses	*Fondante des Charneux* Large fruit, of uneven outline. Flesh very tender and richly flavoured	November	Stand. Espa.	E	W	N
Beurré Capiaumont *Beurré de Capiaumont* *Beurré aurore*	An excellent autumn pear	Oct. & Nov.	Stand. Espa.	E	W	N Graft upon pear stock
Duchesse d'Angoulême *Poire de Pézenas*	Well known by this name	Oct. & Nov.	Stand. Espa.	E	W	N Dry ground; must be on pear stock; against south wall

Name	Season	Form	Aspect	N/S	Remarks
Baronne de Mello *Adèle Saint-Denis* A fine-flavoured large fruit	Oct. & Nov.	Stand. Espa.	E	W	N	Pear stock; does well standard or dwarf
Bon-chrétien Napoléon *Poire lilard* *Poire médaille* *Poire melon* *Captif de Sainte-Hélène* *Bonaparte* *Napoléon* Does best against a south wall, but will do well on quince as a standard	Nov. & Dec.	Stand. Espa.	E	W		Graft upon pear stock; a fine late pear
Colmar d'Arenberg *Colmar d'Arenberg* Graft on pear stock	Oct. & Nov.	Stand. Espa.	E	W	N	A good-looking, but coarse fruit
Triomphe de Jodoigne *Triomphe de Jodoigne* Fine handsome fruit	Nov. & Dec.	Stand. Espa.	E	W	N	Best from quince stock
Van Mons de Léon Leclerc *Van Mons Léon Leclerc* A remarkably fine pear; fruits well	November	Stand. Espa.	E	W		Graft upon pear stock
Beurré Diel *Beurré magnifique* *Beurré royal* *Beurré des trois tours* *Beurré incomparable* Very large fruit; great bearer, but coarse; a wall or standard, in midland counties	Nov. & Dec.	Stand. Espa.	E	W		Graft upon pear stock
Délices d'Hardempont *Poire pomme*	Nov. & Dec.	Stand. Espa.	E	W	N	A very good pear; ripens well
Bergamote crassane *Crasane* A very fine old pear	Nov. & Dec.	Espalier	E	W		Tree is not a good bearer
Beurré Clairgeau *Beurré Clairgeau* A handsome showy pear	Nov. & Dec.	Stand. Espa.	E	W		Graft upon pear stock
Figue *Figue d'Alençon*	Nov. & Dec.	Stand. Espa.	E	W		Graft on quince
Beurré Passe Colmar *Passe Colmar* A very fine pear	Nov. & Dec.	Espalier	E	W	N	Graft upon pear stock; wall; worthless from standard
Saint-Germain d'hiver blanc *St. Germain* Worthless, except for its great size	Nov. to Jan.	Espalier	E	W	S	Wall necessary; used for cooking, baking, &c.
Saint-Germain d'hiver gris	Nov. to Jan.	Espalier	E	W	S	
Beurré d'Arenberg *Beurré Lombard* *Beurré de Cambron* An excellent pear	Jan. & Mar.	Stand. Espa.	E	W	N	Wall, except on quince

NAME OF THE VARIETY AND SYNONYMES.	DESCRIPTION, OR ENGLISH NAME.	TIME WHEN FRUIT RIPENS.	TRAINING.	ASPECT.			OBSERVATIONS.
Beurré gris d'hiver nouveau. *Beurré de Luçon*	Beurré gris d'hiver	Jan. & Feb.	Stand. Espa.	E	W	N	Wall, and very hot situation
Bergamote de la Pentecôte. *Doyenné d'hiver*	Easter Beurré. Does well on quince for standard bushes	Jan. to Mar.	Stand. Espa.	E	W	N	Graft upon pear stock; requires a hot wall
Beurré de Rans. *Beurré de Noirchein Bon-chrétien de Rans Hardempont de printemps*	Beurré de Rans. A very valuable late pear, in a warm situation; bears well	Feb. & Mar.	Espalier	E	W		Pear stock; requires a wall
Bergamote Esperen	Bergamot Esperen. Very good late pear	Feb. & Mar.	Stand. Espa.	E	W	N	Requires a wall
Doyenné d'Alençon *Doyenné d'hiver nouveau*	Easter Beurré	Feb. & Mar.	Stand. Espa.	E	S	N	
Colmar Van Mons	Colmar Van Mons. A very inferior pear	Nov. to Jan.	Stand. Espa.	E	S		
*Messire Jean. *Chaulis*	John Dory. Juicy; good bearer	Nov. & Dec.	Stand. Espa.	E	W	N	
*Catillac. *Poire de tirre Gros Guillot Grand Monarque*	Catillac. The best cooking pear known	January	Stand. Espa.	E			Graft upon pear stock
*Martin-Sec. *Roussélet d'hiver*	Martin Sec. An excellent stewing pear	January	Stand. Espa.	E	W	S	Standard, on pear stock
*Bon-chrétien d'hiver	Bon-chrétien d'hiver. Stewing pear	Jan. to Mar.	Espalier	E			Wall, in a hot situation
*Belle Angevine. *Bolivar Royale d'Angleterre*	Undoile Saint-Germain. Very large	Feb. & Mar.	Stand. Espa.	E	W	S	Wall, on pear stock

* Those marked with an Asterisk are Cooking Pears.

THE PEAR.

The following is a list of pears selected from a great variety grown by the editor. They are all suitable for the midland and north-midland counties; they bear well and ripen to perfection, which is not the case with many of the French and Flemish pears. The late-ripening pears depend much on climate and situation.

English, or Name best known.	Time of Ripening.	Standards on Pear Stock.	Dwarfs on Quince.	Observations.
Aston Town	October	Pear Stock	On Quince	Small fruit, great bearer
Ananas (Beurré d'Ananas)	October	Pear	Quince	Large fruit, excellent
Beurré Bose	November	Pear		Large, excellent
Beurré Brown	October	Pear		Requires a wall
Beurré d'Amanlis	September	Pear	Quince	First-rate, profitable
Beurré de Capiaumont	October	Pear	Quince	Very great bearer
Beurré Rance	May	Pear		Wall pear, south
Beurré Superfin	October	Pear	Quince	Fine dessert pear
Beurré Bachelier	Dec. & Jan.	Pear	Quince	Fine pear, great bearer
Beurré Langelier	December	Pear	Quince	An excellent pear in a warm season
Brougham	November	Pear	Quince	Large and good
Captif de Sainte-Hélène	December	Pear	Quince	Good bearer
Colmar d'Eté	September	Pear	Quince	Good bearer
Citron des Carmes	July	Pear	Quince	Small, one of the best
Comte de Lamy	October	Pear	Quince	Most excellent
Doyenné d'Eté	July	Pear	Quince	Good, the earliest summer pear
Franc real Summer	September	Pear	Quince	Profitable, good
Fondante Van Mons	November	Pear	Quince	Good bearer
Hessel	September	Pear		Profitable, excellent
Joséphine de Malines	February	Pear. South Wall		Very fine late pear
Jargonelle	August	Pear. Wall	Quince, standard	Excellent
Louise Bonne de Jersey	October	Pear	Quince	Good bearer, excellent
Maria Louisa	October	Pear. Wall	Quince	Fine, south wall
Monarch (Knight's)	January	Pear		Good bearer, south wall
Napoleon	November	Pear	Quince	Does not ripen on pear stock
Nelis, Winter	January	Pear. Wall		Small, excellent
Passe Colmar Doré	December	Pear. Wall		Large, great bearer, good
Seckel	September	Pear. Standard or Wall		Small, but fine flavoured
Swan's Egg	November	Pear		An old and excellent variety
Van Mons Léon Seilee	October	Pear. Wall		Great bearer
Williams's Bon-chretien	September	Pear. Wall	Quince, dwrf.	A greatly esteemed pear, but does not keep long

TRAINING AND PRUNING OF PEAR TREES.

The pear is cultivated both as a standard and espalier. The forms in which it may be trained are various; but what we shall now consider are the following: full standards, the pyramid or cone, the vase or goblet, the column, the double contra-espalier in vertical cordon, and tall standards; for espaliers, the Verrier palmette, the simple oblique cordon, and vertical cordon. These forms are the most simple and easily obtained; they are suitable for the places most frequently assigned to the pear, and trees submitted to these forms are durable and fertile.

Training of a Pear Tree in Pyramidal Form.

In this operation we shall consider separately the formation of the wood and of the fruit-branches.

Formation of the Wood.—A tree trained in this form (fig. 35) is composed of a vertical stem furnished from the top downwards, to a foot above the soil, with lateral branches, the length of which should increase down to the base. Each lateral branch should have an interval of twelve inches between it and the branch immediately above, allowing the light to penetrate between them. They should be kept free from bifurcations and furnished with fruit-branches alone, from end to end, and form an angle with the horizon of 25 degrees at the most. In general, the widest diameter of the pyramid should equal the third part of its height; if the height be six

Fig. 35.—Pyramidal Pear Tree.

yards, then the diameter should be two yards at its widest part.

In soils of medium fertility, trees of this form and size are planted ten or eleven feet apart, so that the light may act equally upon their entire circumference.

Young trees are not submitted to their first pruning until the second year after they have been planted. If performed earlier, the pruning takes away too many of the branches, and the quantity of leaves that they should develop is too much diminished. As it is the leaves which cause the roots to extend themselves, these are less developed, and the growth of buds, which the early pruning was intended to promote, will be feeble, poor, and insufficient to form a foundation for the wood-branches of the tree. When the pruning is not performed until the following year, the tree forms new roots, and when the greater part of the branches come to be pruned away, the sap which the roots now supply in great abundance, reacts with force upon the buds that are reserved upon the tree, and a greater length of wood is obtained, during a single summer, than would be obtained in two years by the former method. Time is gained, and the tree is in a more favourable condition for giving the desired direction and growth to the wood. Nevertheless, as the roots of young trees are more or less damaged by deplantation, it is necessary to cut away a portion of the wood in order to re-establish a proper equilibrium with the roots. The suppression of a third of the length of the most vigorous branches will generally be sufficient.

This rule applies to all fruit trees, except peaches, of

THE PEAR. 55

which we shall speak further on. There is no exception save in the very rare case when the tree has been transplanted with *all its roots entire*, and the roots have not been dried in any degree by the action of the air before being again placed in the earth. In this case only should we apply the first pruning the same year that the tree is transplanted.

First Pruning.—This operation is intended to promote the development of the first lateral branches, which ought to grow from the stem, from about a foot above the ground. In order that these branches be

Fig. 36.—Two Years' Graft. First Pruning.

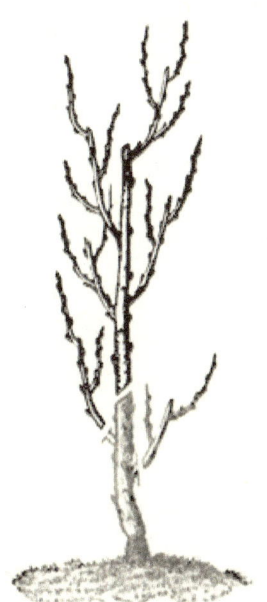
Fig. 37.—Three Years' Graft. First Pruning.

sufficiently vigorous, especially those growing at the base, it will be necessary to develop only six or eight at a time. For this purpose the stem of the young tree is cut at about 20 inches above the ground, at A, fig. 36. The terminal bud reserved at the top should be on the side opposite to that on which the graft has been placed upon the stock at B, in order to maintain the perpendicular direction of the stem.

This mode applies to young trees taken from the nursery, whether of two or three years' grafting, as shown by figures 36 and 37.

In the latter instance, whatever lateral branches have grown upon the stem below the cut are taken off very close to the stem, leaving always the small foundation of the branch upon the stem.

If, while the young trees have stood in the nursery, they have received such attention that the base of the stem is already provided with a sufficient number of lateral branches (fig. 38), such as we would wish to obtain as the results of the first pruning, we apply to them the operations described further on for the second pruning, but always one year after transplantation. It is essential to guard against their bearing fruit at this early period, or they will be debilitated.

During the summer which follows the first pruning, all the buds develop vigorously. When they have obtained a length of four or five inches, we disbud, that is, cut off all the buds from the lower part of the stem to a foot from the ground. From among the buds situated above that point, we reserve six or more of those most regularly placed, but only one at each point. The terminal bud must be maintained in its place by means of a "tutor," or small support fixed at the top of the stem.

Watch carefully that the lateral buds maintain an

THE PEAR. 57

equal degree of vigour. If one of them has grown too long, as shown A (fig. 39), its vegetation must be checked by pinching, that is, by taking off about two-

Fig. 38.—Pyramidal Pear.
Second Pruning.

Fig. 39.—Stem of Pyramidal Pear.
Pinching.

thirds of an inch from the extremity with the finger nails.

Second Pruning.—By spring of the following year the young trees present the appearance of fig. 38. The second pruning is to determine the formation of a new

D 3

series of lateral branches, and to promote the extension of those previously obtained. The new branches should be as numerous as those of the preceding year, and commence about twelve inches above the first. We obtain this result by cutting the terminal shoot about sixteen inches (fig. 38) above its spring. We choose again for the terminal bud one on the opposite side to that from which the branch springs that it grows upon.

The lateral branches previously obtained must be pruned in such a manner as to transform them finally into fruit-branches. But only so much of the branches must be cut away as is necessary to obtain that result, otherwise it will diminish too much the vigour required by the branches to secure their continued growth. The buds which spring from these lateral branches develop themselves too vigorously if not sufficiently pruned, and will with difficulty be put into fruit-bearing condition. The degree of pruning to be applied to them must vary according to the position of the branches upon the stem of the tree. Those near to the ground must be left longer, in order to favour their development. Thus, towards the base, only a third part of the length must be pruned off, and half of those next above them, and finally, three-fourths of the length of those highest up on the tree. Figure 38 explains this operation.

The bud over which we effect the section of the lateral branch should be upon the outside of the tree (A, fig. 40), in order that the shoot which springs from it may follow naturally in the oblique ascent of the other branches. The only exception to this is, when the proper bud is situated too close to the neighbouring ones right and left. In this case select a bud a little sideways of the one that would otherwise be chosen.

If during the preceding summer some of the lateral branches have been imperfectly developed, which takes place occasionally towards the base of the tree, it will be necessary to prune them but little, or leave them entire, in order to impart to them their proper vigour. If these branches be only half the length of the others, it will be well to make an incision on the outside of the main stem, immediately above the spring of the

Fig. 40.—Terminal Bud to prolong a Lateral Branch.

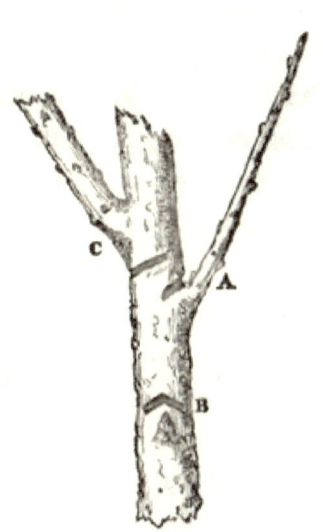

Fig. 41.—Incisions.

branch A, as shown at A, fig. 41. The incision should penetrate the *exterior* wood of the tree, sever the sap vessels which press on that side, and thus compel the sap to develop the branch. The cut should be made with a fine hand-saw, in order that it may heal up less rapidly. If the bud on which we reckoned to form one of the lateral branches remains dormant, the incision will be still more indispensable (B, fig. 41).

When, on the contrary, a branch shall have acquired, notwithstanding pinching, a disproportionate development, it must be cut shorter than the others; if the difference in size between it and the other branches be excessive, an incision must be made like the one shown at C (fig. 41), immediately below its point of attachment to the stem. This will greatly diminish the action of the sap.

During the summer which follows the second pruning, apply the operation of disbudding to the terminal branch, the same as that made upon the young stem during the first year. This must be done so as to leave only six or eight well-placed buds, for forming a second series of lateral branches. Also pinch off the herbaceous extremities of the terminal buds upon the side branches, in order to maintain between them an equal degree of vigour. Be careful that the lateral buds towards the end of the branch do not become too vigorous, and overgrow the terminal one, which ought always to maintain its superiority.

Third Pruning.—In the following spring the tree presents the aspect of the figure 42.

The terminal branch extension is cut in the same proportion as in the preceding year. The extensions of the lateral branches of two years' growth are cut back also in the same proportion as before. The lateral branches developed during the preceding summer are cut shorter, in order to favour the growth of the lower branches. It must be borne in mind that these directions may be modified by particular circumstances, indicated at the time of the second pruning, and that the use of incisions must be resorted to in the cases before specified.

Fig. 42.—Third Pruning of the Pyramidal Pear.

The summer operations are the same as after the second year's pruning.

Fourth Pruning.—Figure 43 indicates the changes that the tree has experienced during the preceding summer.

The fourth pruning differs from the others in several particulars. We allow to the new extensions of the lower branches only half the length of former pruning, because they are on the point of attaining the limit beyond which they must not be allowed to grow; besides which, they have now attained a size which will enable them to maintain their proper vigour.

We allow to the new branch extensions of the second series two-thirds of their length, and suppress the half or three-fourths of the length of the upper branches. These various ramifications are cut a little

Fig. 43.—Fourth Pruning of the Pyramidal Pear.

longer than before, because the lower branches have now less need of protection, and it is time to commence giving to the tree its pyramidal form. The new terminal shoot is treated the same as in preceding years.

During the following summer, we observe the same

treatment as in preceding ones; but as the lower branches have now almost attained their proper length, it will be necessary to restrain the further growth by

Fig. 44.—Fifth Pruning of the Pyramidal Pear.

pinching off the terminal shoots that have acquired a length of ten or twelve inches. The sap will, by this means, be forced back, to the benefit of the higher parts of the tree.

Fifth Pruning.—The tree now shows somewhat of

its intended proportions (fig. 44), the lower branches being inclined a little by their own weight, the tree assumes its pyramidal shape. The pruning does not differ from preceding years, only that the inferior branches having now attained their intended length, we cut their new extensions quite close. The other branch extensions should be all cut following the line A B (fig. 44). The summer operations are the same as the preceding year.

Sixth Pruning.—This does not differ from the fifth, only as the lateral branches lengthen they increase in weight, and hang down too much towards the ground on the neighbouring branches, and must be brought into their first direction again, by means of strings or supports, so that the space between each may always remain equal. The same treatment is carried on until the twelfth year, at which time the tree presents the aspect of figure 35.

If there is still sufficient earth for the roots to extend themselves further, the tree will still have a tendency to increase its size, and this may be turned to advantage. Allow the terminal shoots and the lateral branches to extend themselves afresh, but in such a manner as to preserve the proportion between the height and the diameter, as before directed.

OBTAINING AND MAINTAINING FRUIT-BRANCHES.

All that we have hitherto said has related to the formation of the wood of the tree. We shall now describe the operations that promote the growth of fruit-branches, and the care and treatment they require.

The fruit-branches of *pépin* fruits that have been regularly pruned every year, ought now to be growing upon the entire length of every branch without interruptions. In open standards the fruit-branches ought to occupy the entire circumference of the tree. In espaliers the only part without them is that against the wall. The fruit-branches are generally constituted towards the end of the third year following their first development. If this result is obtained earlier, it indicates a diseased or enfeebled condition in the part of the tree where they grow.

The fruit-branches are kept as short as possible, that the fruits may be close to the principal branches; they will then receive the most direct action of the sap, and become larger than if placed at a further distance from its source. We shall now explain how these different results are to be obtained.

First Year.—Fruit-branches are developed from the less vigorous buds upon the wood-branches. In order to obtain a continued series of fruit-buds upon the entire length of a branch extension, it is necessary to cut back a little of the branch, otherwise the wood-buds on one part, towards the base, will remain undeveloped. We have already given directions, p. 39, as to the length that the branch extensions should be cut back according to their degree of inclination.

Suppose the pruning has been duly performed upon the branch extension (fig. 45), by the beginning of May the branch will be covered with buds upon its entire length (fig. 46). The vigour of the buds will be greatest as they approach the highest part of the branch, and those quite at the extremity will, unless arrested, acquire a great development. Now, it is only

Fig. 45.—Wood Branch Extension.

the weak buds that become fruit-spurs; it is, therefore, important to diminish their vigour. This result is

Fig. 46.—Wood-Branch Extension.

obtained by pinching. As soon as the buds intended to form fruit-branches have attained a length of about four inches, they must be pinched off with the nails (fig. 47). In performing this operation some pinch off too much, leaving two or three leaves only towards the base (fig. 48). Two inconveniences may follow: very soon the fragment of the shoot ceases to grow, and after the fall of the leaves all that remains is a small stump without any buds (fig. 49), which dries up and

Fig. 47.— Bud rightly pinched.

dies, leaving a vacant place the following year. This occurs most frequently upon certain varieties of pears, which do not produce any eyes near the base of the

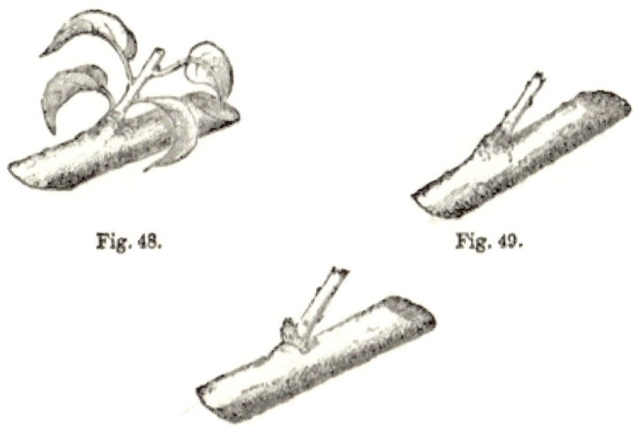

Fig. 48.   Fig. 49.

Fig. 50.
Buds pinched too far back, and the Result in different Stages of Growth.

branches; such as the Bon-chrétien d'Hiver, the Beurré Magnifique or Beurré Diel, the Doyenné, &c. Sometimes there appear, a year or two after this excessive pinching, two buds placed on each side the lower end of the suppressed shoot (fig. 50), which, in three years more, become flower-buds. The vacant part is then filled up, but at least a year is lost in the formation of flower-buds. At other times, when the lower leaves of the suppressed shoot have eyes at their base, those eyes give place to so many premature buds immediately after this excessive pinching (fig. 51). These premature shoots do not become well-constituted branches, and set for fruit less freely than branches

Fig. 51.

produced from shoots pinched in the proper manner, that is, leaving to them a length of two or three inches (fig. 47).

Each of the branch extensions of the wood is furnished with a bud so favourably situated, as regards the action of the sap (A, fig. 45), that the repeated pinchings to which we may submit the shoot which it produces, diminish its vigour very slightly, and it always produces a too vigorous shoot. It will be better to treat such in the following manner:—When it has obtained a length of about three inches, cut it off at the base, leaving only a very small portion of the lower end (A, fig. 53). The two supplementary buds, which accompany the primary one (C, fig. 52), give place almost immediately to two small shoots much less strong than the one suppressed (fig. 53). We pinch off the most vigorous of the two, and the one that is reserved (which must also be pinched if necessary) gives place to a small branch which easily sets for fruit.

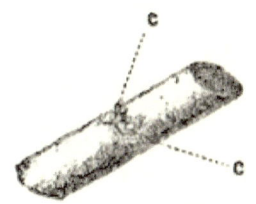

Fig. 52.—Side Buds.

Fig. 53.—Side Buds after removal of principal Bud.

Fig. 54.—Pinched Shoot with new Shoot D.

The first pinching is generally sufficient to arrest the too rapid growth of the shoots. The most vigorous, however, will often produce an anticipative bud towards their summit (D, fig. 54). These also must be pinched when they have attained a length of three or four inches.

If some shoots have been neglected until they have attained a length of eight or twelve inches or more, it will then be too late to pinch them. If, indeed, we then cut them off at four inches from their base, we should see all the eyes situated at the base of the leaves, that we wished to form into flower-buds, develop themselves under the action of the sap, which has set in to flow to that part, into anticipative buds, which all at once find themselves in too strait bounds. It will be better, therefore, to submit these neglected buds to *twisting*, at about four inches from their base, B A, as shown at fig. 55, and to pinch off their tops. It will

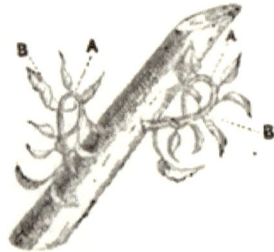
Fig. 55.—Shoot submitted to Twisting.

Fig. 56.—Small Shoot near the lower end of the Extension.

follow from this double operation that the development of these buds will be arrested, and the eyes at the base will advance without pushing forward into anticipative buds.

Such is the treatment required by shoots intended to become fruit-branches, during the summer of their first development. It is seen that the whole of these operations are not to be practised at once. The proper time for each is indicated by the different stages of growth, and this treatment should be followed during nearly the entire period of vegetation.

Second Year.—In consequence of the various operations that we have described, the shoots which grow upon the branch extensions (take for example figs. 45 and 46) have given place to a series of small branches, which are less vigorous as they approach the lower end of the branch. During the succeeding winter, a mode of treatment must be pursued to be regulated by the degree of vigour of the shoots, the end in view being to fatigue and enfeeble the branches, and by this means hasten their fruit-bearing.

The buds situated towards the lower third of the extension (fig. 46) have grown slightly and become very short branches, resembling fig. 56. We perform

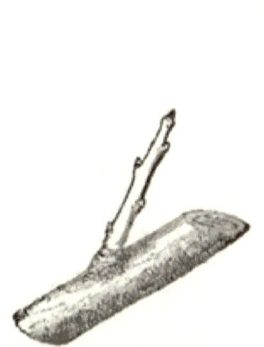

Fig. 57.—Small Shoot in the middle of Extension.

Fig. 58.—Pinched Branch.

no operation upon these; they will transform themselves into fruit-branches.

The shoots upon the middle third portion of the extension (fig. 46) are rather more grown, and resemble fig. 57, and are called *dards* (darts). Nothing further is to be done to them.

Lastly, towards the upper third part of the extension (fig. 46) the shoots have pushed out with more vigour, but they have been submitted to pinching or twisting. These have given place to the following series of shoots :—The less vigorous, or of medium vigour, resemble those of fig. 58. These are broken off at A, about three inches from their base, and immediately below a bud. The fracture fatigues the branch by producing a contused and torn place. There is then less probability that the lower buds will develop into vigorous shoots; the small piece left above the upper bud will still further favour the set to fruit by allowing the sap to expend part of its action upon the issue at the end of the branch.

The other more vigorous branches, that have been repeatedly pinched during the summer, now resemble fig. 59. These should receive the *partial fracture* (B), as shown in the figure. If broken completely, the sap, being more abundant than in the other shoots, would be confined in too narrow limits, and make the lower buds, that should set for fruit, push out into vigorous shoots. The partial fracture affords a sufficient issue for the sap, while enough is retained to enable the lower buds to develop a rosette of leaves.

The shoots that have been twisted during the preceding summer, present the appearance of fig. 60. They must be submitted to the complete fracture at A, if they are of less or medium vigour, or to A and complete at B if very vigorous.

The branches we have now spoken of are the only ones that ought to be found upon the extension, fig. 46, if the operations of pinching and twisting have been duly performed in the preceding summer. But it may happen that some of the shoots have been neglected, and grown to a length of twelve or eighteen inches, and are more or less thick. These shoots are called *brindilles*. If let alone, these may set for fruit; but as they spring from the summit of the extension, their situation is unfavourable to their development; besides, such long fruit-branches produce confusion in

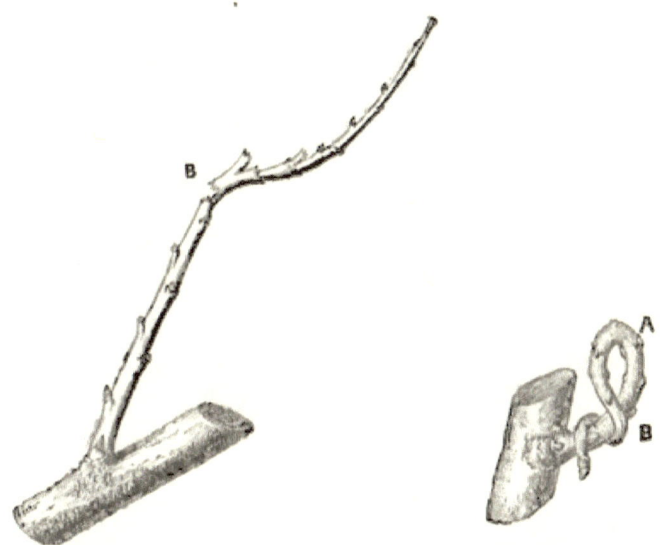

Fig. 59.—Branch pinched several Times. Fig. 60.—Twisted Branch.

the tree, and should be shortened. They may be broken completely at four inches from their base, at C, fig. 61, if of feeble or medium vigour; if more vigorous, they may be completely broken eight inches from their base, and partially at four inches (fig. 62). If these branches possess extraordinary vigour, they assume the character of *gourmand branches*, and may readily be transformed into fruit-branches by placing a Girardin side graft at their base, and cutting the branch off above the graft at the winter pruning.

Third Year.—During the summer that has followed these operations, and in consequence of them, the branches have produced the following results:—

The minute branches situated near the base of the

Fig. 61.—Twig of medium Vigour.   Fig. 62.—Vigorous Twig fractured Twice.

extension (fig. 56) have only developed a rosette of leaves, having a bud in the centre, and have increased a very little in length. Their appearance after vegetation, as shown by fig. 63, is that of a bud very thick at the upper part. This bud will expand its flowers in spring. These minute branches, which are in the third year of their formation, now become fruit-spurs.*

The *dards* (fig. 57) have developed two or three very short buds, which have produced minute branches, as shown at fig. 64. So also have the branches submitted

* French name, *lambourdes*.

to complete or partial fracture (figs. 58, 59, 60, 61, and 62); two or three of their buds have pushed

Fig. 63.—Small Branch transformed into a Fruit-Spur.

Fig. 64.—Small Two Years' Shoot.

forward into small shoots, and produced other small branches extremely minute (figs. 65, 66, and 67).

Fig. 65. Shoot completely Fractured.

Fig. 66.—Shoot partially Fractured.

If, during the summer, the bud situated towards the end of the branch has grown into a slightly vigorous shoot, it must be pinched down to four inches (A, fig. 67). There is no further operation to be applied to these various productions during the second winter.

Fourth Year.—During the third summer, the fruit-spur, shown at fig. 63, has fructified. It has formed,

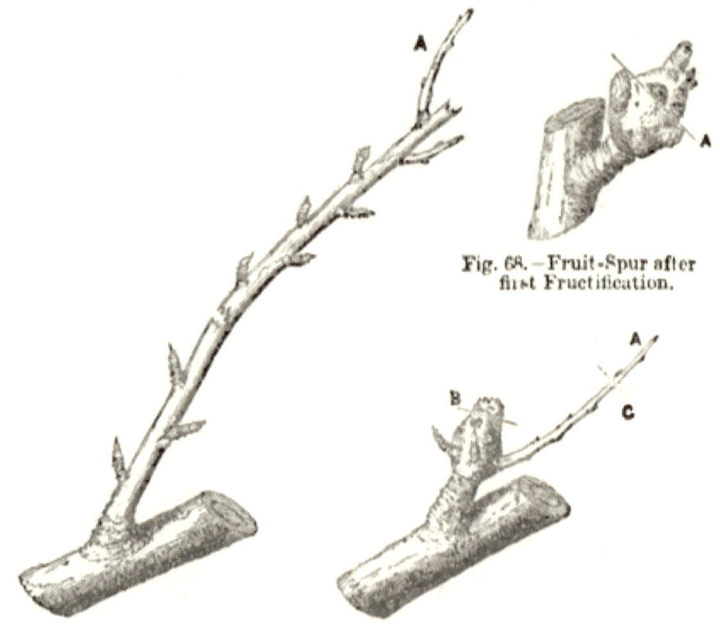

Fig. 67.—Shoot doubly Fractured.   Fig. 69.—Fruit-Spur with small Branch.

Fig. 68.—Fruit-Spur after first Fructification.

at the point where the fruit grew, and its accompanying rosette of leaves, a kind of spongy swelling, shown at figs. 68 and 69, called a purse (*bourse*). We observe, besides, certain buds borne upon very short branches springing from the leaf-roots of the purse. These become flower-buds in the course of two or three years. Sometimes one of the eyes growing at the leaf-roots develops itself into a vigorous shoot. It must be pinched to four inches. The small branch, which is the result of this (A, fig. 69), must then receive the complete fracture at C. The only further care to be applied to the purses consists in cutting off the top, which is now in a state of decomposition, at A (fig. 68), or at B (fig. 69).

The *dards* have extended their small branches and grown a little longer; the small flower-bud at the end is about to expand, and will give place to a *purse*, like fig. 68. The same treatment must be observed as

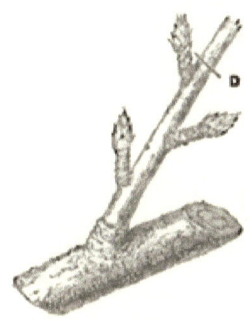

Fig. 70.—Small Three Years' Branch, bearing Fruit-Spurs.   Fig. 71.—Small Branch Two Years after Fracture, bearing Fruit-Spurs.

directed in the former case, at the period of winter pruning.

The fractured branches (fig. 65) also bear flower-buds (fig. 71). The time has arrived to cut off at D, fig. 71, the little extension left at the end of the branch. The *purses* that they will produce must also receive the treatment before described. Lastly, the branches submitted to partial fracture (figs. 66 and 67) also bear small fruit-spurs (figs. 72 and 73). These branches may be cut back at A, for the flower-buds being formed, there is no longer danger of the sap being restrained in too confined limits, nor of the branches pushing out into vigorous but unproductive shoots.

Maintaining the Fruit-Branches in Bearing Condition.—As we have already stated, the fruit-spurs (fig. 68), which have fructified, may produce new flower-buds two or three years afterwards, the fruit-spur having ramified, as shown at fig. 74. So also will each of the small fruit-spurs situated upon the branches of which

Fig. 72.—Two Years' Branch after partial Fracture, bearing Fruit-Spurs.

Fig. 73.— Two Years' Branch after double Fracture, bearing Fruit-Spurs.

we have spoken above. In six years after their first fructification, each of these fruit-spurs will be constituted as shown at fig. 75. If the fruit-spurs have not been injured during their development, and the tree is sufficiently vigorous, they will, after a certain time, present the appearance of fig. 76. The question now arises, ought we to allow them to continue to increase in this manner to an indefinite extent? If so, the fruits will soon grow at too great a distance from their principal branch, and receive an insufficient supply of sap, the action of which will be further restrained by the numerous little ramifications it will have to traverse before reaching the fruit. Besides, such a development of fruit-spurs produces much con-

Fig. 74.—Six Years' Fruit-Spur. Fig. 75.—Eight or Ten Years' Fruit-Spur.

fusion throughout the tree, prevents the light from penetrating, and confines fructification to the ends of the outside branches. If the fruit-spurs are allowed to grow in this way it will be necessary to greatly diminish the number of wood-branches, and leave large vacancies in the tree.

It will thus be seen that it is necessary to restrain the fruit-spurs within reasonable limits, not allowing them to exceed a length of more than about two inches. When, therefore, they have attained the dimensions, as shown at fig. 75, they must be cut back at A. The action of the sap will thus be forced back towards the base, and give rise to new fruit-buds.

If the fruit-spurs have already been allowed to attain too great dimensions (fig. 76), it will be necessary to reduce them very gradually; they must be cut successively, first at B, the year following at C, and so on. If cut immediately at D, the action of the sap being too much restrained, the fruit-spurs will develop vigorous shoots and become transformed into wood-branches.

Fig. 76.—Method of Pruning an Old Fruit-Spur.

Such are the operations required to develop the fruit-bearing functions of *pépin* fruits, and to maintain them in a healthful and fruit-bearing condition. We have shown that their fertility is the result of successive mutilations of the wood of the lateral branches, by which their vigour has been restrained and diminished. It must not, however, be lost sight of, that this object has also been powerfully promoted by allowing the annual extensions of the lateral branches to grow to a greater length, thus opening a large outlet for the sap, which thus acts with less intensity upon each of the separate shoots. The pruning, almost always too short, that is applied to these extensions, has the contrary effect, giving rise to shoots of such extreme vigour, as to require five or six years of successive mutilations to transform them into fruit-bearers.

Attention to the Fruit.—To complete the preceding observations, it should be observed, that nothing tends more to exhaust a tree and destroy the fruit-spurs than allowing too great a quantity of fruit to ripen upon it. It absorbs almost all the sap, and not only is the formation of new fruit-buds prevented for the following year, but the existing ones are destroyed for want of nourishment. The principal branches put out only small, poor terminal shoots, and the roots are scarcely able to put forth sufficient force to extend their ramifications, so as to draw nourishment from a wider zone of earth beyond that which has been impoverished by the precedent vegetation. The tree then remains in a languishing and sterile condition for years. The end which nature purposes to attain by the fructification of fruit trees differs from the object that we have in view. Nature only seeks the production of the largest quantity of seeds, irrespective of the pulp of the fruit, in order to promote in the greatest degree the multiplication of the individual species. The object we strive to attain is the largest quantity of pulpy material, without regard to seeds. The quantity of seeds is in proportion to the number of the fruits; the larger the number, the less pulpy are they, and their quality is, in an equal degree, deteriorated.

It is therefore of the greatest importance to suppress the superabundant fruits, in order to regulate the crop and improve the quality. We lose in number, truly, but have an equal weight, for the fruit retained increases by means of the sap of those suppressed. As to the proportion of the fruit to be retained, the following rule should be observed:—The number of fruits allowed to ripen should equal about the fourth of the number of all the fruit-branches. The suppression must only be made when nature has made her choice, that is, when the fruits have attained about one-fourth of their development.

Training of the Pear in Vase or Goblet Form.

Trees in the pyramidal form are, in some situations, liable to injury from high winds. When that is the case, the vase or goblet form may be substituted. But it is not otherwise to be preferred, for it requires as much room as the pyramid form, and does not present so great a fruit-bearing surface.

Trees in vase form should have a diameter of about six feet six inches (and an equal height), so that the solar rays may act upon the whole interior surface of the vase. An interval of twelve inches should be left between each of the branches. Supposing the tree to be 20 feet in circumference, there should be about twenty branches at the base, from which to form the tree.

The branches may either be trained vertically, or made to cross each other alternately right to left, following an angle of 30 degrees, as shown in fig. 77. We consider the latter form preferable. The sap acts more equally throughout the entire extent of the branches, which also fruit more regularly, and the tree can better support itself when completely formed.

The method of proceeding to develop the wood is as follows:—Choose plants that have been grafted a year, and apply the first pruning; when they have been

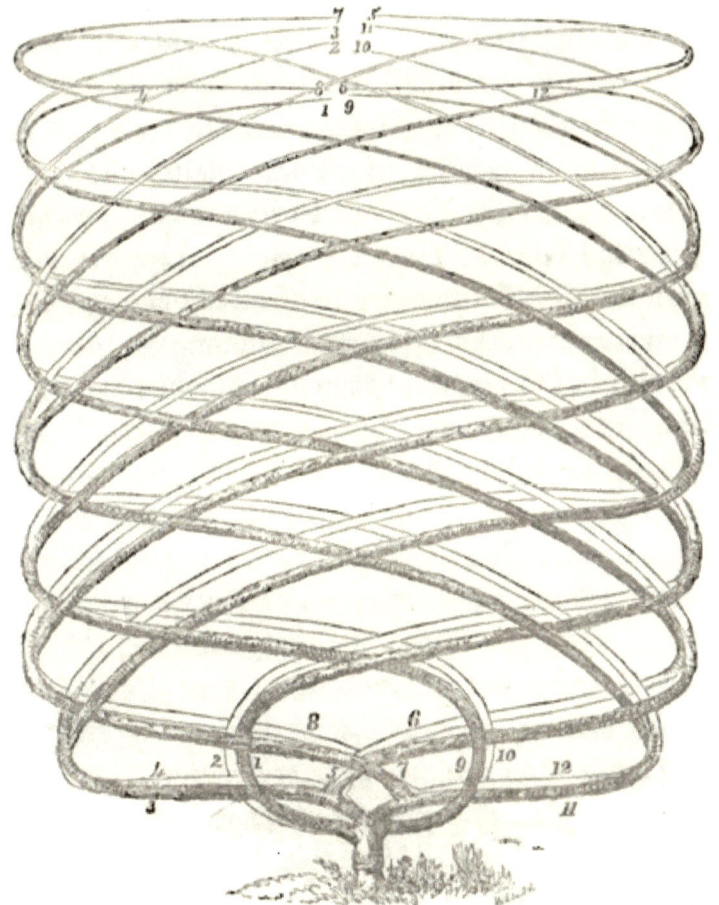

Fig. 77.—Pear trained in Vase Shape.

planted out another year, cut them down to 16 inches from the ground. During the summer select five shoots, and maintain an equal degree of vigour by pinching.

At the second pruning cut off each of the branches to 16 inches from their base, above two buds growing laterally, so as to make each of the branches fork; lower the branches a little, and dispose them regularly round the circumference of the stem by means of a hoop. During the summer equalise the vigour between the ten shoots that have now been obtained. At the third pruning, cut back each of the ten shoots to 12 inches from their base, to make them fork a second time. Incline the branches again, and equalise the spaces between them by means of two hoops, the uppermost being the larger. Treat the twenty shoots that have grown during the summer in the same manner as the previous ones. At the fourth pruning, suppress only the third of the length of the new branches, and again incline them downwards to an angle of about twenty degrees, then raise the ends of the branches in a vertical position at about three feet from the stem, and keep them in that position by means of additional hoops. During the summer allow only one terminal bud to develop. When the time arrives for the fifth operation, cross the branches at the place of their second forking, directing them alternately right and left, inclined to an angle of thirty degrees. Figure 78 shows the plan of a pear tree trained in this form, and how the branches should be crossed. The new extensions obtained during the previous summer must be left entire, and so on from year to year, until the tree has attained its proper size. The inclined position of the branches will cause them to put forth numerous shoots, each of which must be trained in a spiral direction, to be arrested only when the tree attains a height of about six or seven feet, when it will be fully formed, and resemble the figure 77.

As the tree increases in height, each of the branches must be *grafted together* by approach at each of the

Fig. 78.—Plan of Fig. 77.

points where they cross. This will give great strength and solidity to the tree, and enable it to dispense with any other support when the wood is completely established.

The fruit-branches, which are not shown on the figures, are formed and kept in bearing in the same manner as in pyramid trees.

Training of the Pear Tree in the form of a Column.

The two forms we have described are the most convenient, especially the pyramid. They both, however, require much space, and are therefore less suited for small gardens, as so few varieties can be cultivated. The column form will be better adapted for small gardens. This form consists of a single vertical stem, growing to the height of 20 feet or more, and furnished from bottom to top with fruit-branches.

The column form is not so sightly, but presents several advantages; it casts less shade, and occupies less ground than pyramid trees, permits the cultivation of other products in the neighbourhood, and more varieties may be grown upon the same space of ground. We are disposed to think, moreover, that the fruit-branches form more readily, and being better exposed to the action of the sun, and growing directly from the main stem, are more fertile, and the fruit is generally finer. But this success can only be obtained under certain conditions. First, the trees must be grafted upon quince stocks. If grafted upon the pear they will become too vigorous, and produce only wood-branches. The trees should be planted in warm, light soil, of medium fertility.

The method of forming these trees is very simple. At the first pruning the young stem is cut much longer than for pyramid trees. The object aimed at is to cover the entire length of the stem with *dards* or twigs, and on no account with vigorous woody branches.

During the summer all the buds may be allowed to develop themselves freely, preserving the pre-eminence to the terminal shoot. At the second pruning, the new extension is treated in the same manner; the most vigorous wood-branches of the new extension developed a little below the point of attachment are cut close off; the more feeble branches, and the twigs below them, are broken above a well-formed bud three inches from their spring. Finally, the twigs at the base are left entire. During the summer they must be left to grow without interference, except to protect the terminal shoot.

At the third pruning the small twigs that have grown upon the lower part of the boughs cut during the preceding year must be broken, if more vigorous than twigs; the same must be done to all the other shoots that are too vigorously developed below this point. The new ramifications growing upon the extension pruned the preceding year must be treated the same as the former ones, and so on year after year. The re-

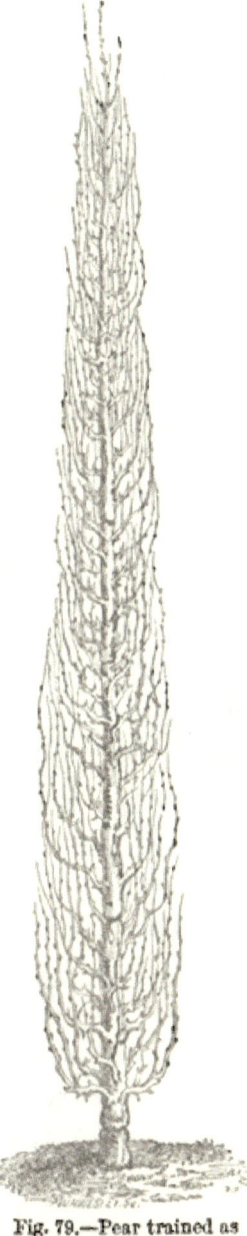

Fig. 79.—Pear trained as Column.

sult of this mode of treatment, and of the conditions laid down, will be a quick formation of small branches, extending from the base to the summit of the stem, and covered with fruit-spurs, also some wood-shoots, &c.; these latter must be cut off every year above their foundations. As these little branches will become very numerous, and produce confusion, some of them must be cut back every year, but only here and there, for fear of concentrating too much the action of the sap upon too small a space, and thus injuring the formation of fruit-spurs.

M. Choppin completes his series of operations by applying to the stem a number of annular incisions, with the view of restraining the flow of sap in the lower part of the tree, and diminishing its too great vigour. The first incision is made about ten inches above the graft, towards the fourth year of pruning; the other incisions must be applied from time to time as required. The more vigorous the tree, the more frequent and numerous the incisions.

That which most characterises this mode of pruning is, that the summer operations of pinching and disbudding are never practised. All the shoots are allowed to develop freely. If the more vigorous of them were pinched off, the sap not having, as in the preceding forms, a great space to run through, would be checked in its course, and cause the small branchlets to shoot out and become vigorous woody branches, which would otherwise have remained simple twigs.

Training of Pear Trees in Vertical Cordon as Double Contra Espaliers.

Of the forms that we have already described for standard pear trees, the pyramid or cone will generally be found most suitable; the goblet form affords less fruit for the same surface of ground occupied, and is only suited for very hot, dry soils. The vase or goblet form certainly offers some advantages; but it has also grave inconveniences, amongst which are the following:—

1. The wood is not completely formed until towards the twelfth year, and fruit does not appear until about the fourteenth year after plantation.

2. It requires great space, and is unsuitable for small gardens.

3. The training requires such great care and precise observation, that few gardeners are found competent for the task.

4. It is almost impossible to protect the tree from spring frosts.

5. Lastly, there is an insufficient proportion between the product of fruit and the ground that the tree occupies.

The form we shall now describe presents none of these disadvantages.

The double espaliers must be placed upon the ordinary borders of the fruit-gardens; the borders should be about two yards wide, having a road between each border of about a yard in width, running, as nearly as possible, north and south. The figures 81, 82, 83 show the details.

EXPLANATION OF FIGURES 81, 82, AND 83.

A. Posts placed 18 or 20 feet apart.
B. Galvanised iron wire.
P. Galvanised iron wire, forming a line between the two posts, holding them together, and made fast to a wall.
D. Screw tightener.
E. Laths placed in front to support the stems of the trees.
F. Laths placed upon the back surface.
O. Wires crossing the tops of the posts at right angles made fast to walls.
N. Apple trees trained in low horizontal lines.

The posts, A, should be made of resinous wood; if passed through a solution of sulphate of copper it will add to their durability; their length should be 9 feet, and about 5 inches diameter. They should be sunk in the ground to a depth of 20 inches in the middle of the garden border, and be from 6 to 7 yards apart from each other. The galvanised iron wire (P, figs. 81 and 82) passes over the top of each post through a ring or staple, and is fastened at each extremity to a wall.*

Other galvanised wires (O, figs. 82 and 83) also pass over the tops of the posts, but in a direction at right angles with the first, and are also fixed to the top of the walls. These wires are made tight by the screw

---

* If there are no walls to fasten the supporting wires to, the posts must be somewhat stronger, and the lower post left thick and sunk deeper in the ground.

(D, figs. 81 and 82). The posts are thus solidly fixed from the top to the bottom. Then extend from the front and back of the two posts other iron wires (B, figs. 81, 82, and 83), traversing a ring on the sides of the posts. These wires have also each a screw tightener. Finally, fix upon these last four wires (B) on each side, a series of wood laths about an inch and a half wide, and a third of an inch thick (E F, figs. 81, 82, and 83). Fasten the laths by a knot of fine wire, 12 inches apart upon the surface of the wires back and front, alternately, as shown at fig. 82.

Next proceed to the planting. The trees must be planted back and front, a tree against each of the laths to be trained as single vertical cordon espaliers (fig. 80). In the front of each of the borders, a line of apple trees may be planted, to be trained in low horizontal espaliers (N, figs. 83 and 112.)

Let us compare, for a moment, the results of this new mode of training with the results of the old methods. Suppose two fruit-gardens, both having exactly the same extent of surface, and both equally divided into borders six feet wide, separated by roads 40 inches in width, which is the common arrangement for plantations of pyramidal trees. Let one of these gardens be set apart for pyramids, and the other for double contra espaliers. The total length of wood-branches that we can obtain from pyramids may be set down at 2,500 yards, and the contra espaliers will give a length of 5,000 yards, double that of the pyramids on the same extent of border.

It must be added to this that the maximum product

could not be obtained from the pyramidal trees until about the fourteenth year after planting, while the contra espaliers would begin to bear at the sixth year at latest.

These contra espaliers give, on the same quantity of ground, twice the extent of wood, and therefore twice the quantity of fruit compared with the other forms of training, and attain their maximum product of fruit eight years sooner.

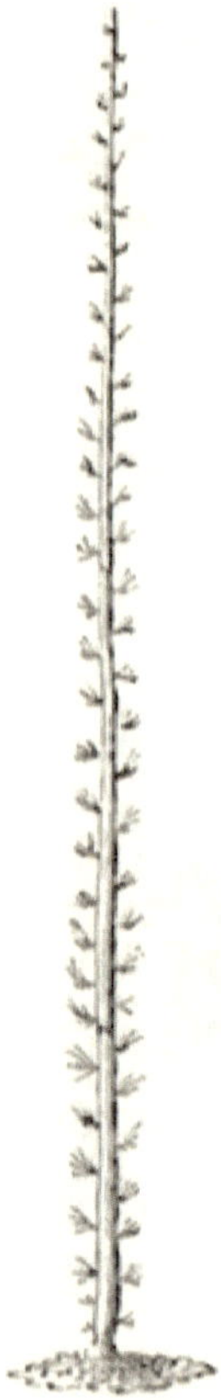

It may be objected, however, that the new method will be more expensive than the old forms of training. This is true, because, by the old method, only one tree would be required for a border, while twenty must be planted on the new method. But it will be a sufficient answer to this objection that three years' produce would more than repay the extra expense, and there would still remain, to the advantage of the new method, five years of more than double the crop of ordinary trees before the others came into full bearing. Besides, the expense may be diminished about one-half; instead of purchasing ready-grafted trees, buy young stocks that may be had for a mere trifle, and plant them out in the nursery; after they have been grafted a year, plant them in their proper positions upon the border.

Fig. 80.—Pear Cordon.

By following this course, the period of maximum fruit-bearing will be retarded two years, but there will

THE PEAR.   93

still remain six years to the advantage of the new method.

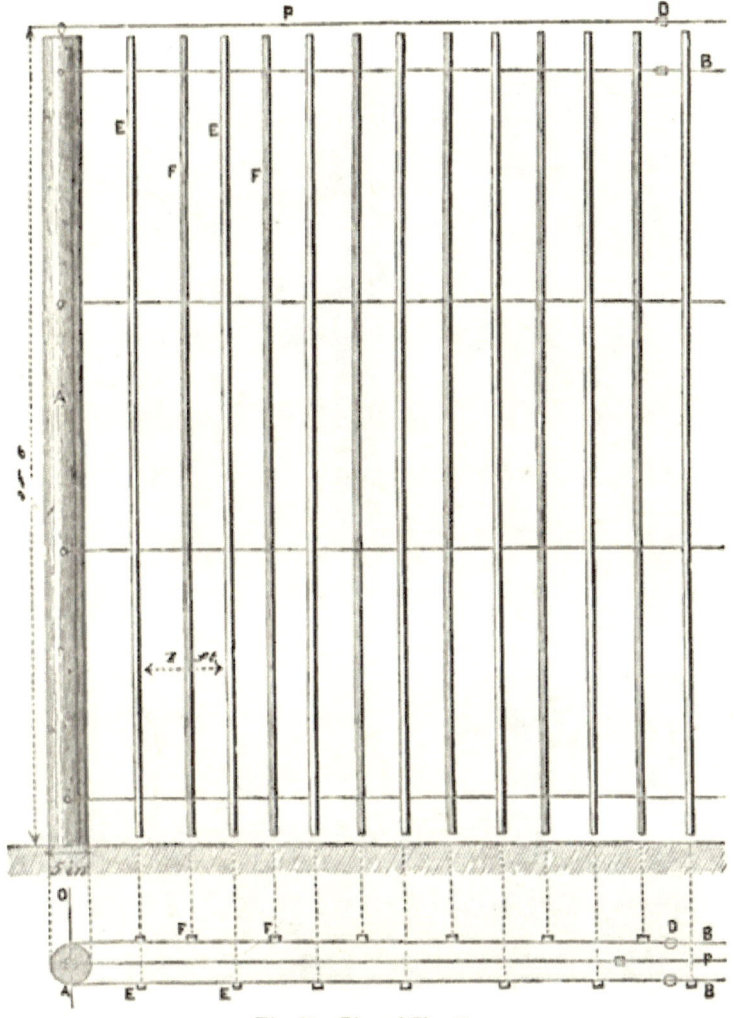

Fig. 81.—Elevation of Contra Espalier.

Fig. 82.—Plan of Fig. 81.

On a comparison of the two methods the whole matter stands thus :—

1. The maximum product of fruit is obtained eight years earlier.

94 FRUIT TREES.

2. Double the amount of produce is obtained from the same extent of ground.

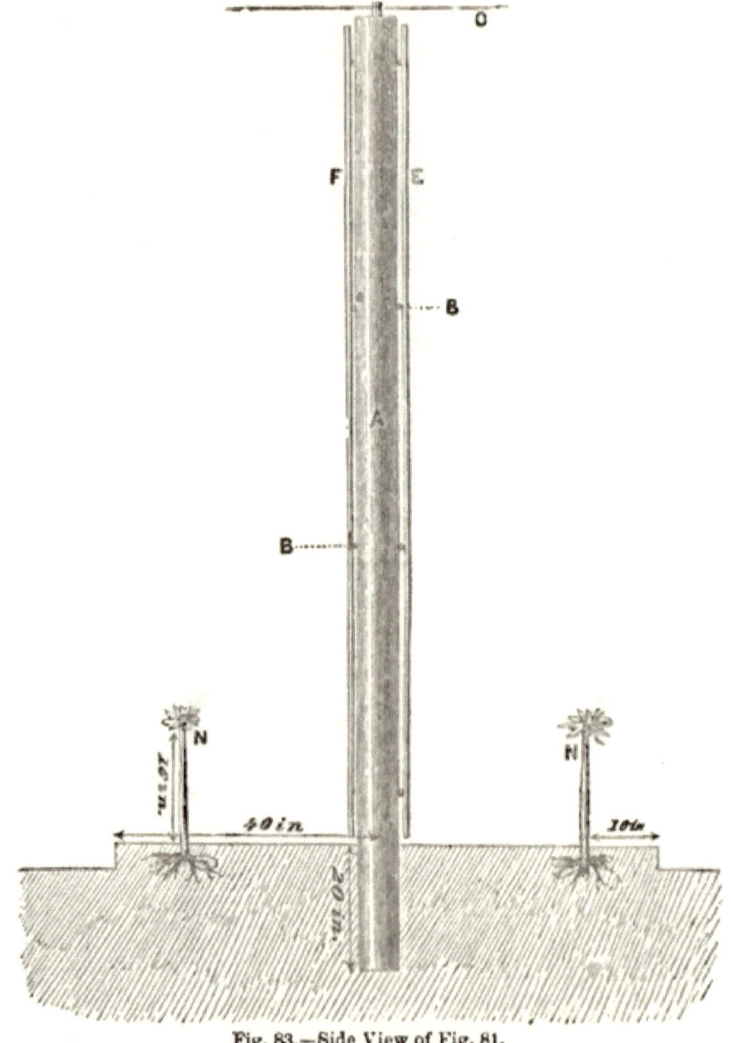

Fig. 83.—Side View of Fig. 81.

3. The easy way in which they may be protected from the late frosts of spring. For this purpose it is sufficient to place a white cloth over the espaliers from end to end, which may remain until the end of May.

4. The wood-branches are more regularly exposed to the action of the light, and better furnished with fruit-branches.

5. The advantage of being able to plant a greater number of varieties in a small garden, and of prolonging the duration of fruit-bearing.

6. Great simplicity in the training and pruning operations.

7. Lastly, the vacant spaces left by the decay of some trees may be more easily filled up than in the case of cone or other forms of trees.

With these advantages we do not hesitate to recommend almost the exclusive adoption of the new method for pear trees in the open air. The mode of forming the wood of these trees is, in all respects, the same as that described further on (p. 119) for espaliers in vertical cordons.

Training of Pear Trees in the Form of Verrier's Palmette.

There are many forms in which espaliers and double espaliers may be trained, but the most convenient are those known by the name of *palmettes*.

These are very simple and easy to manage, and accommodate themselves to walls of various heights. The best form of palmette is one we saw for the first time at the district farm of Saulsaye, under the management of M. Verrier, whose name we have given to it.

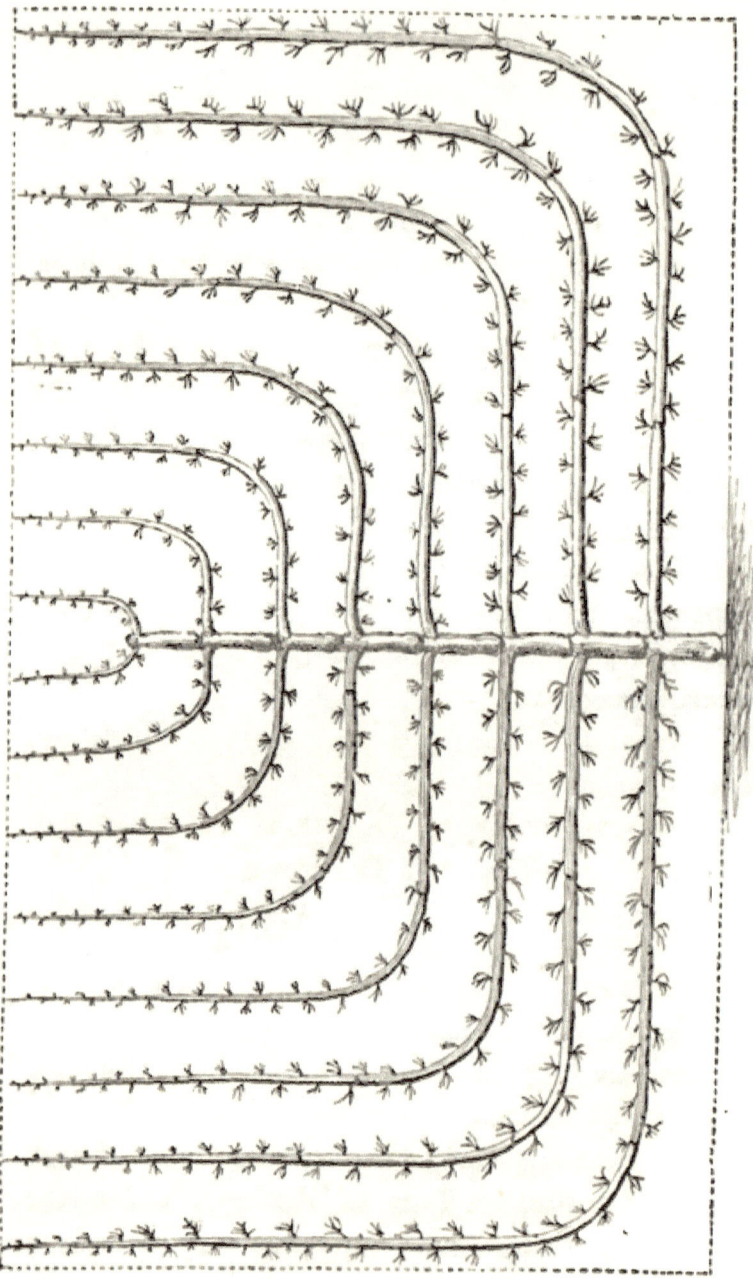

Fig. 84.—Verrier's Palmette.

Trees submitted to this form of training (fig. 84) are composed of a vertical stem, with a series of branches growing from the sides opposite each other at 12 inches apart.

These branches are at first trained horizontally, then turned upwards in a vertical direction, and rise to the summit of the wall.

This form is preferable on some accounts to the *palmette with oblique branches.*

The form is more favourable to the equal distribution of sap throughout the tree, and vegetation is more easily and equally maintained. The method to be followed in obtaining this form is as follows :—

Choose for planting grafts of one year. Plant the trees at such distances against the walls that each will cover a surface of about 20 square yards. Suppress such a proportion of the stem as is required to establish an equilibrium between the tree and its roots.

First Pruning.—Do not apply the first pruning until the trees have well taken root, or rather, until they have been planted one year. Cut the stem at about 12 inches from the ground, A (fig. 85), immediately above three buds, one upon each side. Two of these buds will form the first lateral branches, and the highest or front one the extension of the stem.

During the summer, preserve upon each of the young stems the three buds only that we have just mentioned, and keep them in an equal state of vigour. If one of them grows faster than the other, loosen it from the wall, incline it a little downwards, and raise the feebler one.

Second Pruning.—After the fall of the leaves the trees will resemble fig. 86. Cut back the two branches one-third at A, in order to make them put out fresh shoots, and consequently fruit-branches from end to end. If one is more vigorous than the other, cut it a little shorter, and leave the feeble one rather longer. In pruning espaliers, the cut must always be made

Fig. 85.—First Pruning of Palmette.

Fig. 86.—Second Pruning of Palmette.

above a front bud, in order that the cut part may be directed towards the wall.

Cut the stem at B, about five inches above the spring of the lateral branches, leaving a bud well placed for forming the new extension of the stem. No more lateral branches are allowed to grow during the second year, in order to avoid the risk of weakening those already formed, which will remain feeble if the stem be lengthened too rapidly. Maintain during the following summer an equal degree of vigour between the new extensions of the two side branches.

Third Pruning.—The year following the trees will resemble figure 87. Proceed now in the following manner:—

Cut back the side branches the same as the first year, reducing the new extension one-third. Cut the stem at A, about six inches above the previous one,

Fig. 87.—Third Pruning of Palmette.

and above three well-placed buds for obtaining a new set of laterals during the following summer. From this period a fresh set of side branches may be developed every year, for the lower ones have now acquired sufficient power. Keep the vegetation equal throughout the tree during the summer.

Fourth Pruning.—The figure 88 shows the progress that the tree has made during the preceding vegetation. Cut the new extensions as in previous prunings. Cut also the stem at A, to obtain a new set of lateral

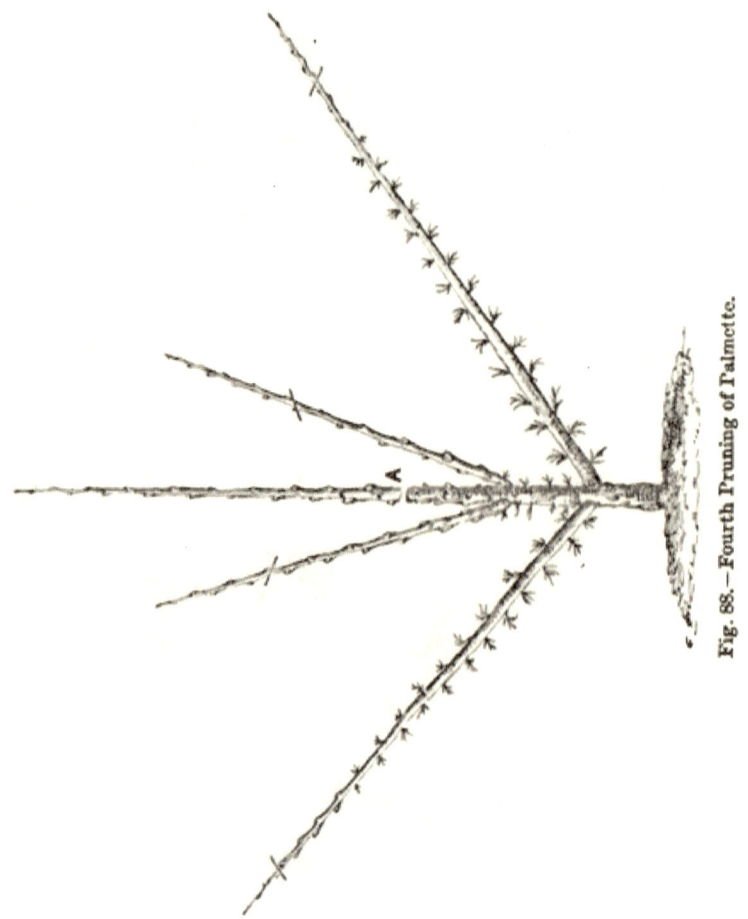

Fig. 88.—Fourth Pruning of Palmette.

branches. Attend to them during the summer as in former years.

Fifth Pruning.—By this time the trees have acquired the development shown by figure 89. Cut the stem at A, to obtain a fourth set of side branches. Cut the

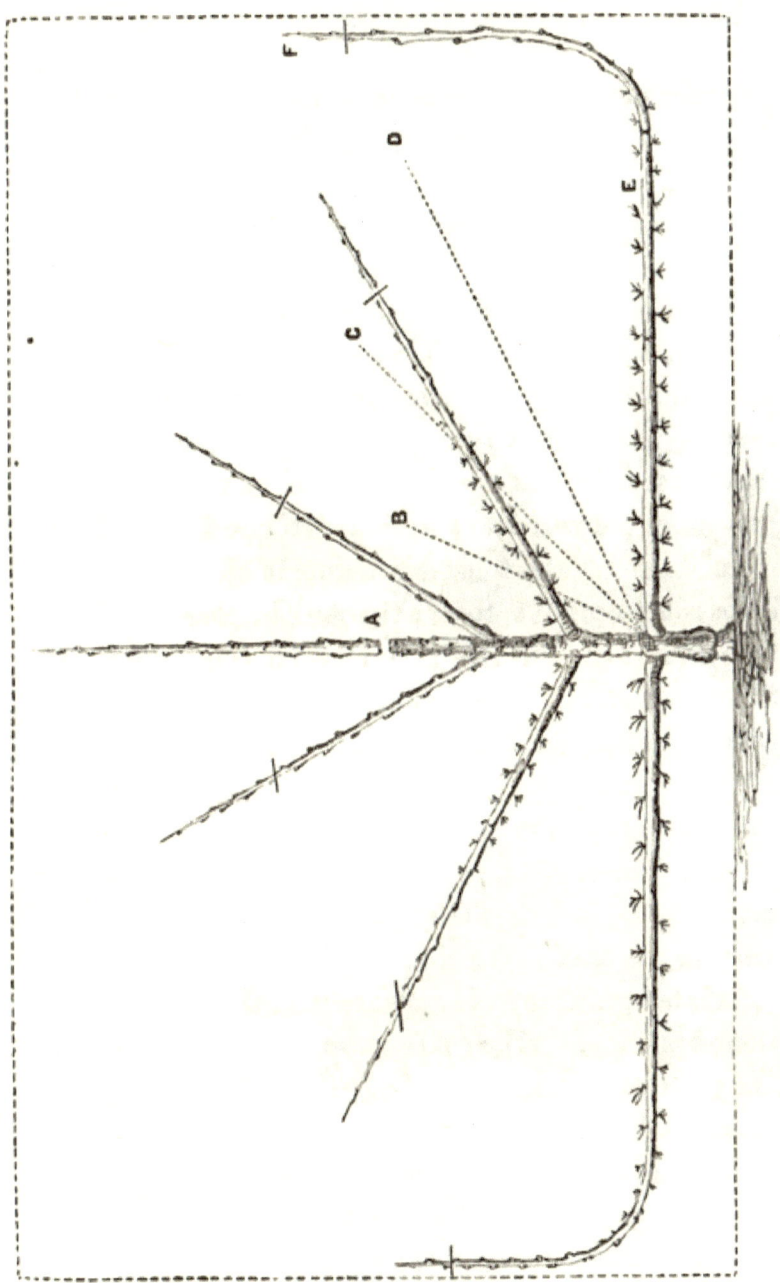

Fig. 89.—Fifth Pruning of Palmette.

extensions of the laterals as in former years. By this time the lower laterals have generally grown sufficiently long to allow of their being curved in a vertical direction, as shown by the figure 89. A bent piece of wood may be used to form the curve and keep the branches in position. They must be encouraged to extend themselves in a vertical direction, forming successive extensions, to be cut back one-third of their length each year. When they have reached the top of the wall they must be cut off to 18 inches each year below the coping, in order to leave room for the terminal shoot, required each year, to draw the sap to that point, and in passing upwards to supply and nourish the fruit-branches. All the lateral branches are subjected to this treatment. Towards the sixteenth or eighteenth year the wood of the tree will be completely established. It will then cover a space of about 20 square yards, and present the aspect of figure 84.

The symmetry and regularity of the wood are important, not only as regards the appearance of the tree, but also, and most of all, to secure an equal degree of vigour in all parts, and thus to promote the fertility and duration of the tree. We shall not always find at the time of winter pruning buds suitably placed for forming the new branches. When this happens, the inconvenience may be counteracted by budding, in August, on the places where the buds are deficient. [If desirable a bud of another variety may be inserted.]

Pruning the Fruit-Branches.—All that we have said at present respecting Verrier's palmette relates to the wood-branches. The treatment of the fruit-branches is the same as that of the pyramidal trees, the only exception being, that in wall trees, the buds nearest to the wall must not be allowed to grow, but be pinched off as they appear.

Nailing up of Wall Trees.

Only the wood-branches of pears trained as wall trees, and their shoots, must be nailed to the wall.

Winter Nailing.—The following rules must be observed:—Train each branch in a perfectly straight line, from the place where it springs from the stem to its farthest extremity. The smallest deviation from the straight line creates an obstacle to the circulation, and branches of excessive vigour will grow out near the curve, and uselessly absorb a great quantity of the sap.

Place the branches which grow at the same height from the opposite side of the stem in exactly the same line of inclination as their fellow-branches on the other side, otherwise the lower will be less vigorous than the higher ones. If, however, the equilibrium of vegetation has been already broken, it will be necessary to lower the strong branches and raise the more feeble ones.

The branches which are finally to occupy an oblique or horizontal position must not be forced in that direction at once, but gradually; if brought into that position suddenly, the sap will all flow up the stem, and the development of the side branches be almost wholly suspended. The branch E (fig. 89) was first inclined to B, then to C, and the following year to D, and only when grown to the length of F was it brought to its final position E. The other side branches must all be inclined in the same gradual manner.

The Summer Nailing.—The summer nailing of pear trees relates only to the new shoots growing from the side branches; these shoots, when they have attained a length of twelve inches, must be nailed to the wall or trellis, in the same direction as the branches upon which they grow.

In nailing to a trellis, a small straight stick may be employed, fixed at the extremity of each branch, and in a parallel line with it, to conduct and support the young shoot.

Two methods of fastening branches to the wall may be used; nailing them with list or fastening to a trellis.

Fig. 90.—Garden Hammer.

Nailing with List is the most perfect and convenient method. The kind of hammer and a convenient form of nail-basket are shown in figures 90 and 91. If the old lists are used again, it is a good precaution to boil them, in order to destroy the eggs of noxious insects with which they abound. It is an advantage to have the walls covered with a thick layer of plaster, so that the nails may be driven in sufficiently deep at any

Fig. 91.—Nail Basket.

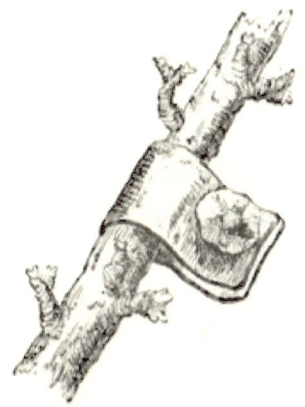

Fig. 92.—Branch fixed against a wall.

point. When this cannot be done, on account of the expense, or for other reasons the wall is unsuitable for nailing, it is best to fasten by means of a trellis.

Fastening by Trellis.—The trellis may be made either

Fig. 93.—Wood Trellis.

of wood or iron wire. There has always been a prejudice against the latter material; but as long experience has proved that no injury results from its use, even in the case of peaches, we do not hesitate to recommend it in preference to wood, as being much the more economical of the two.

The form of the trellis must depend upon the form that is intended to be given to the tree. If made of wood it must be well nailed together, painted with three coats, and fastened with staples.

A wire trellis must be made as follows:—Extend along the wall a sufficient number of lengths of gal-

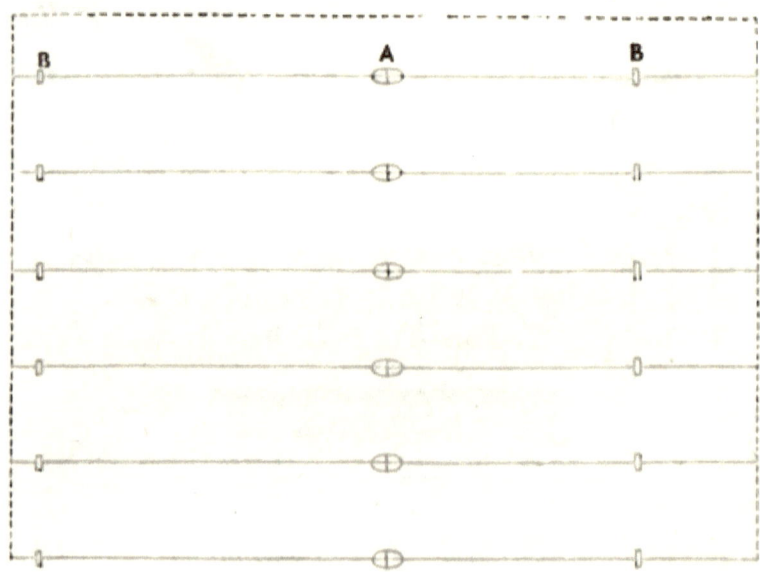

Fig. 94.—Wire Trellis for Palmette.

vanised iron wires, about twelve inches apart. These wires, being securely fastened at each end, must be supported, at distances of a yard apart, by iron pins having a hole through the thick end (B, figs. 94 and 95). The wires must be made as tight as possible by means of a tightener (A, figs. 94 and 96). This tightener is used thus :—When one end of the wire has been made fast to the wall and passed through the eyes of the supporting pins as far as the half of the entire length

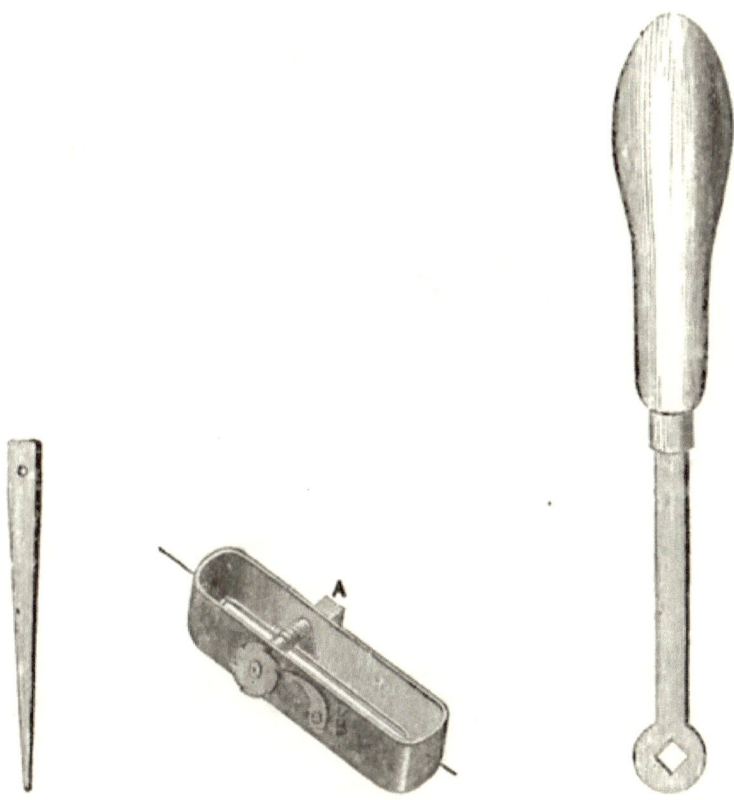

Fig. 95.—Fastener.   Fig. 96.—Tightener.   Fig. 97.—Key.

of the line, it is brought through one end of the tightener, through the centre of the axle, and out at the opposite end, then through the eyes of a number of pins sufficient for the second bolt of the wire, then fasten the wire at the other extremity, and tighten a little. Drive the pins into the wall at their proper distances, then tighten the wires as much as possible, using the key (fig. 97). When sufficiently tight,

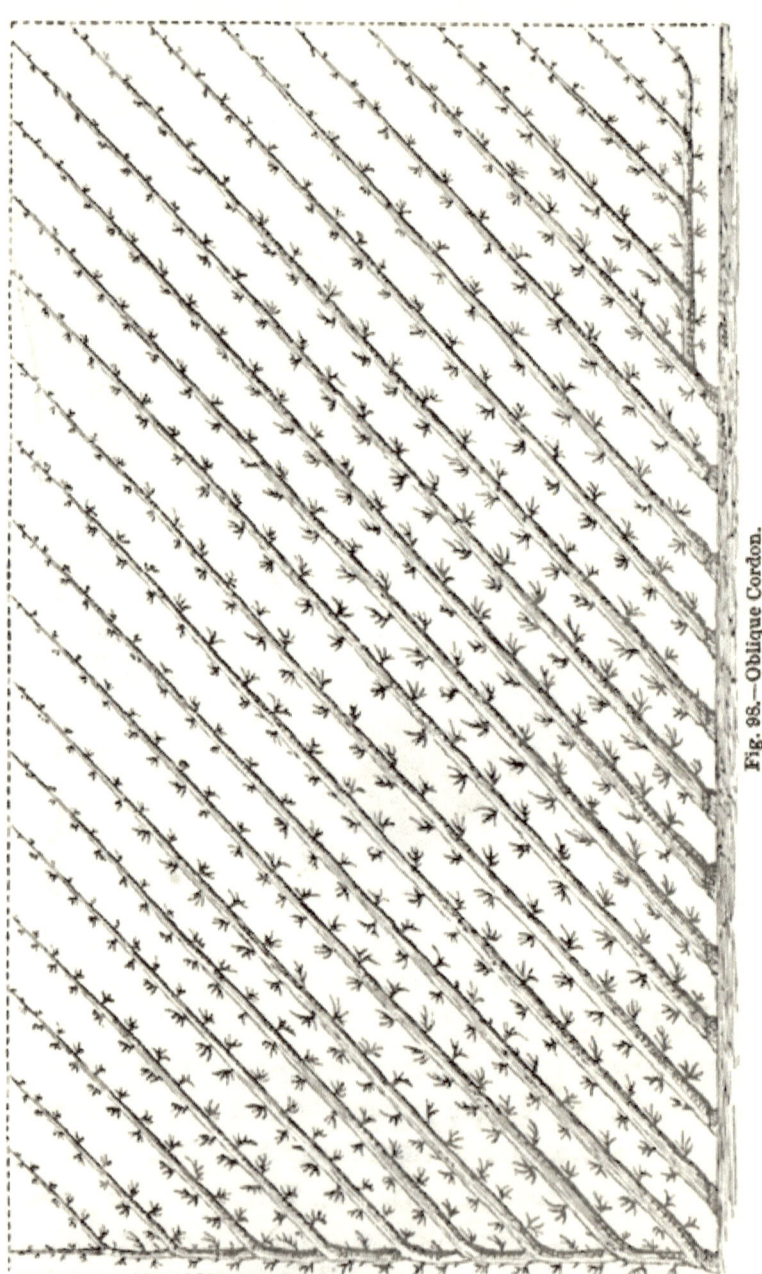

Fig. 98.—Oblique Cordon.

lower the stop upon the toothed wheel (B, fig. 96), which will hold all secure.

The best material for tying the branches to the trellis is osier. It is a security against injury if a small piece of cloth or soft material be placed between the wood of the trellis and the branches, at each place where they are tied. When the trellis is made of wire, the same result may be obtained by twisting together the two ends of osier upon the wire before applying the branch.

It will be necessary during the summer to see that the branches are not injured by the bandages becoming too tight. When this is the case the ligatures must be immediately removed.

Pear Trees in Oblique Cordon.

At present we have only described such methods of training as will result in complete and full-sized wall trees in the course of sixteen or eighteen years, covering a surface of 20 yards and upwards. The care required in obtaining even the most simple of these forms, and the difficulty of maintaining an equal vegetation in all parts of the tree, often render it very difficult for ordinary gardeners to succeed with them.

To remove these inconveniences we have now invented a new and original form, much less difficult to manage, which may be made to cover the entire surface of a wall in a very short time, and to obtain the maximum of product at an early period, without shortening the duration of the trees. We were the first to apply this method, in 1852, and gave it the present name of *oblique cordon* (fig. 98). The following is the way to proceed :—

Choose healthy and vigorous young trees of one year's grafting, carrying only one stem. Plant them

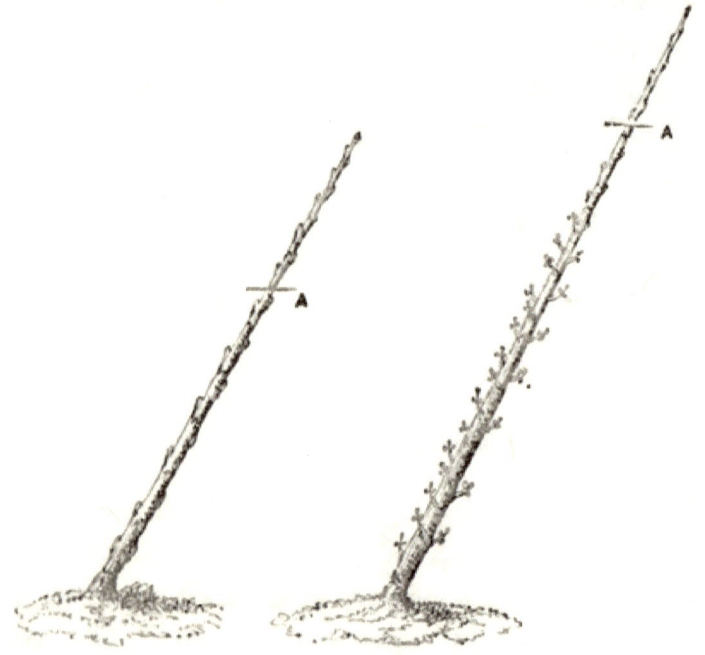

Fig. 99.—Cordon, First Year.   Fig. 100.—Cordon, Second Year.

16 inches apart, and incline them one over the other at an angle of 60 degrees. Cut off about a third of the length at A (fig. 99), just above a front fruit-bud.

During the following summer favour as much as possible the development of the terminal shoot; all the others must be transformed into fruit-branches by the same means as described for pyramidal trees. By spring of the following year each of the young trees will present the aspect of figure 100.

THE PEAR. 111

The second pruning has for its object to transform the lateral shoots into fruit-spurs; the new extension of the stem must be cut back a third. If the terminal extension has grown but slightly, and shows signs of weakness, the cut must be made lower down, on the two-years' wood, in order to obtain a more vigorous

Fig. 101.—Cordon, Third Year.

terminal shoot. Apply the same treatment to these young trees during the summer as during the preceding one ; the result is shown by figure 101.

By the time of the third pruning, the young stem has generally attained two-thirds of its entire length; it must then be inclined to an angle of 45 degrees, following the line B, fig. 101; the same terminal shoot and side branches must be pruned as the last year. If

Fig. 102.—Cordon, Fourth Year.

the stem had been inclined in this manner at the first, the consequence would have been a growing out of vigorous (*gourmand*) branches at the base of the stem, to the injury of the terminal shoot. The new shoots must be treated as usual. Figure 102 shows the state of the tree at the end of this year's vegetation.

There is now nothing more to be done than to complete the tree, by continuing the same treatment until it reaches the top of the wall. Arrived at its final height, it must be cut every year about 16 inches below the coping of the wall, to allow space for the growth of a vigorous shoot every year, which will force the sap to circulate freely through the whole extent of the stem.

If the wall runs east and west, it is not important to which side the stem is inclined; but if the wall extends north and south, it should be inclined to the south, to afford as much light as possible to the underside fruit-branches. When the wall is built on the descent of a hill, the trees should be inclined towards the summit, or their growth will be too soon arrested by the top of the wall. The trees being planted 16 inches apart, and each developing a single stem, the result will be an espalier, composed of a parallel series of slanting trees having a space of about 12 inches between each stem (fig. 98).

If it is desired to fill up the vacant space on the wall left by the inclined position of the trees, a *half palmette* may be grown for that purpose. To accomplish this, first plant a young tree, and treat the same as the others; after it has been inclined to an angle of 45 degrees, allow a strong shoot to grow from the base of the stem, and develop itself freely during the summer. The following year incline it parallel with the other stem, and twelve inches from it. During the summer allow another shoot to grow from the bend of the previous one, incline it as before, and so on each year until the vacant space is filled.

The vacant space at the other end must be filled up thus:—The last tree must be planted at about two yards from the limit of the espalier; it must first be treated as the others have been; then, instead of lowering it to 45 degrees, it must be lowered a little further; the following year lowered still more, and when the stem of the plant has acquired such a length that when placed horizontally it will occupy the whole space shown in figure 98, it is brought into its proper horizontal position. During the following summer allow the four or five shoots, intended to form upward branches growing upon it, to develop themselves.

Wall trees trained in this way attain their full size in five years, a gain of at least ten or twelve years compared with other methods.

By this plan the trees become fruitful in the fourth year, and attain their maximum in the sixth year, while other and larger forms require twenty years to attain their maximum. If the extent of wall is limited, only a small number of varieties can be planted by the ordinary method of growing large trees; while the method we have now described allows of a considerable number of varieties being planted, their fruit ripening throughout the season.

If a large pear tree die, it will require fifteen or eighteen years to replace it. With the new methods of oblique cordon it will be only necessary to proceed thus:—Dig a hole, about 16 inches wide and 20 inches in depth, in the centre of the spot left vacant by the dead tree. Cut away the roots of the neighbouring trees which grow into the hole. Drive two thin pieces of board, about twenty inches square, to the level of the ground, at each end of the hole, plant the tree, a young vigorous one, in the space between them, and cover over with good fresh soil. The pieces of wood will prevent the roots of the trees on each side encroaching on the ground of the new plant. They will soon rot, but the new tree is then in a state to take care of itself. By this means the vacant space will be filled up in five or six years. The wood of such trees as we have now described is more easily established than any other, and the regular inclination given to each stem renders the equal distribution of the sap easy and certain.

Objections have been made to this form. It has been said that the limited extent given to the wood would conduce to such great vigour as to injure the fruit-bearing qualities of the tree. But this vigour is counteracted by the trees being planted so closely together upon such a small surface of ground. It has been also said that trees so near to each other will be unable to live. We answer that an extent of wood is only allowed to each tree in proportion to the extent of soil that the roots occupy. It has been also objected that such a mode of planting is more expensive than the old method. This is true as regards the first expense; but in addition to the operations of pruning being much more easily performed by the new method, it must be considered that ordinary wall trees do not attain their maximum of fruit-bearing until the sixteenth or twentieth year; while the oblique cordon becomes fruitful in four years, and repays its original cost three or four times over. If the walls are sufficiently high—seven feet—nothing can be more simple and profitable than the method we have now described; but if less than that height, it will be better to keep to the palmette form.

Trellis for the Oblique Cordon.—The most simple form is that shown by figure 104. For a wall nine or

Fig. 103.—Iron Pin.

ten feet high, there will be required three cross pieces of wood or strong iron wire, firmly attached to the wall; then a series of laths fastened to the cross pieces, 16 inches apart, and inclined 45 degrees, each of the laths supporting one of the young trees.

The cost will be much less if the trellis be made of wire (fig. 105), as invented by M. Thiry, jun. At the points A, B, C, D, E, F, strong round nails, fig. 103, are driven firmly into the wall; then at the points G, H, I, J, K, L, pins with a hole through the head (fig. 95). The end of the wire is made fast at the point A, then passed through the eyes of the iron pins G H, then supported by the two nails B C, it is passed through the pins I J, under the two nails D E,

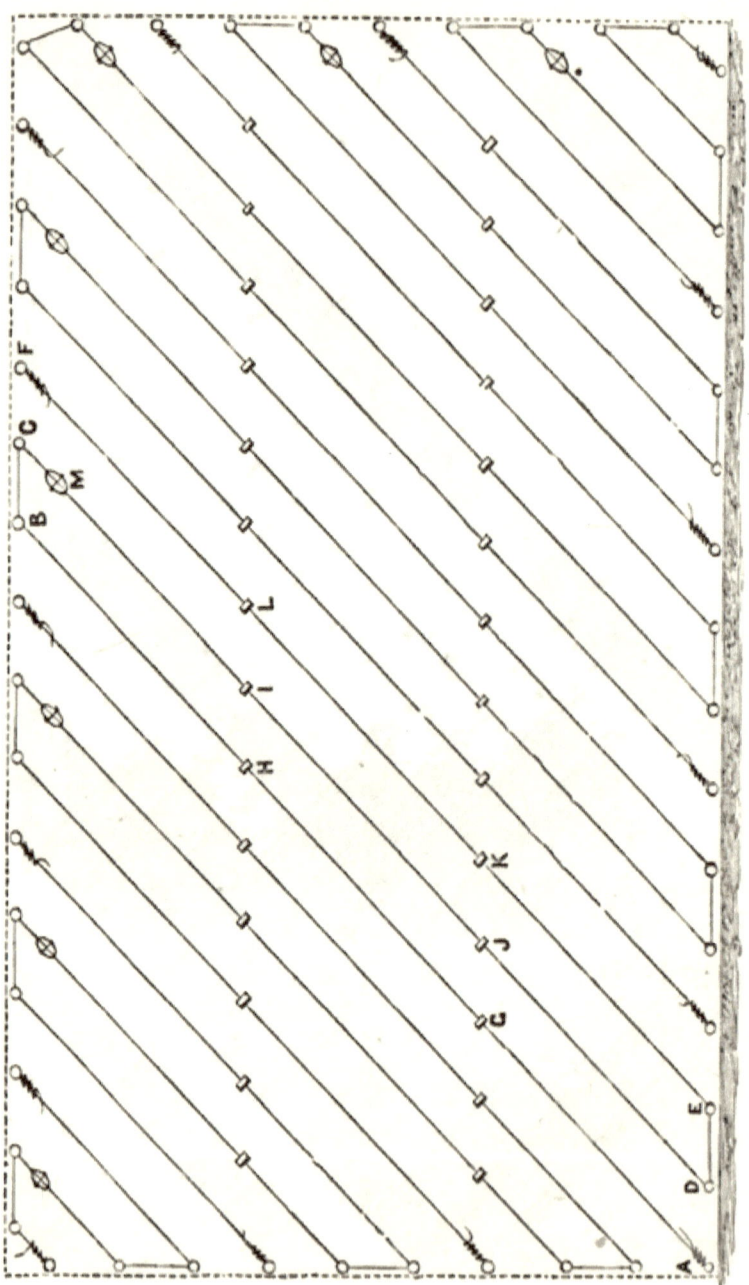

through the pins K L, and then fixed upon the nail F. It is made tight by means of the tightener (fig. 96), fixed below the nail C, as shown in the figure at M.

A little oil placed upon the nails B, C, D, E, will facilitate the tightening of the wires by means of the tightener. The same operation being repeated all along the wall, it will then be covered with a series of well-stretched wires, in parallel lines, inclined 45 degrees, 12 inches apart from each other. This trellis costs (in France) about 5s. per square yard, not including fixing.

Training of the Pear in Vertical Cordon.

The walls are sometimes of such height, as in the case of gable ends of buildings, that it would be inconvenient to train in oblique cordon. Whenever the wall is higher than 13 feet we recommend the vertical cordon to be employed.

The method for planting is, in this case, the same as for the oblique cordon, only the trees are planted vertically (fig. 106), and 12 inches only apart from each other. The stems must be trained towards the summit of the wall, in the same manner as directed for the oblique cordon.

If the wall will not allow of the trees being fastened up with nails and list, recourse must be had to a wood trellis, as shown at fig. 107, or a wire one (fig. 108), which is much less expensive.

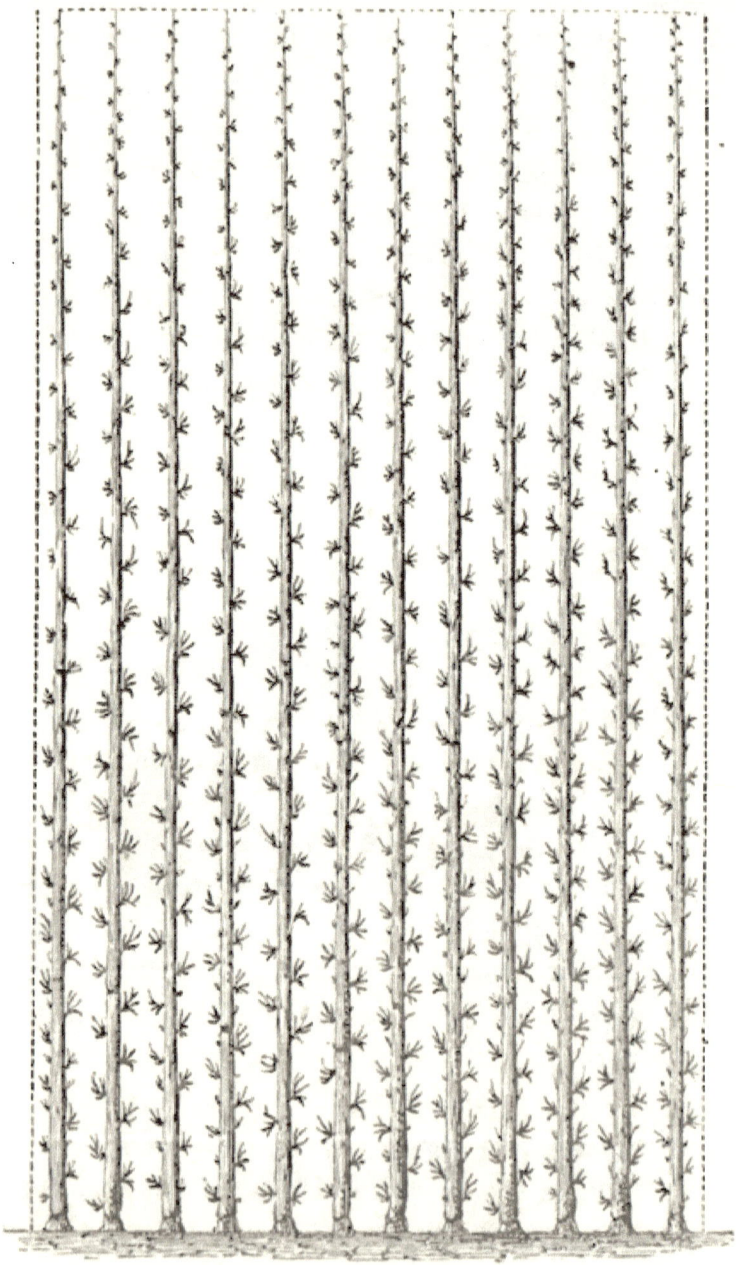

Fig. 106.—Pears in Vertical Cordon.

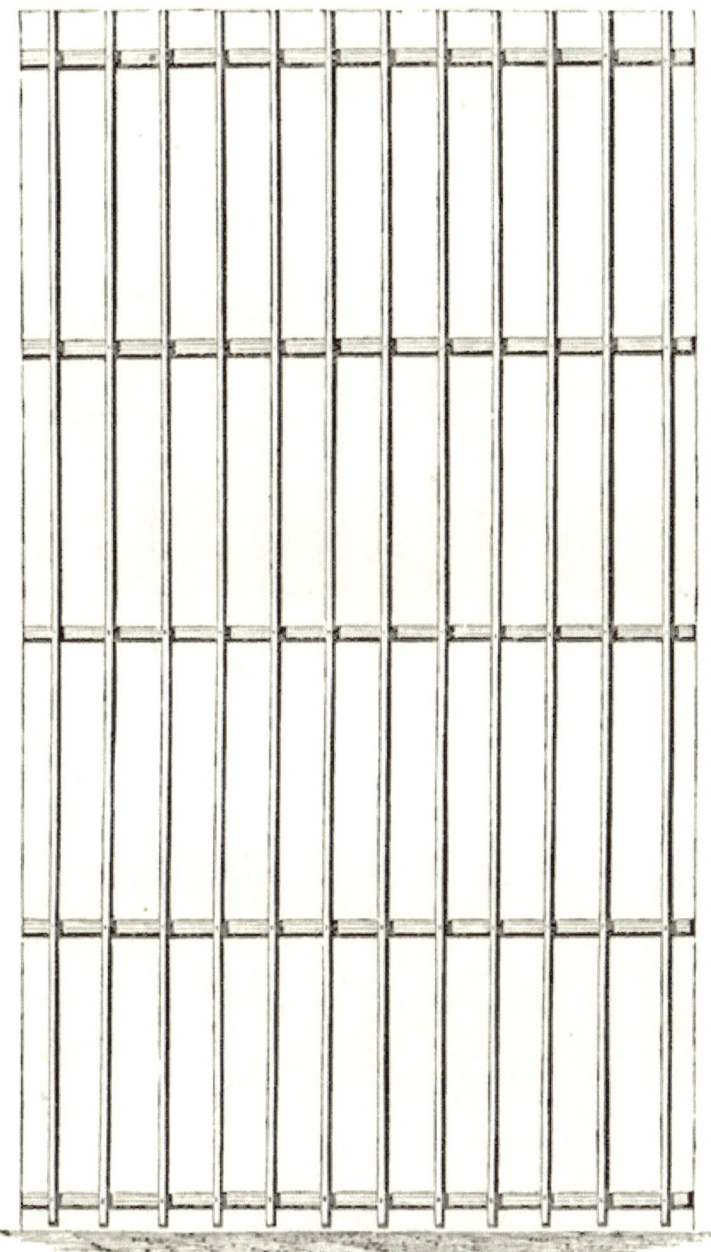

Fig. 107.—Wood Trellis for Vertical Cordon.

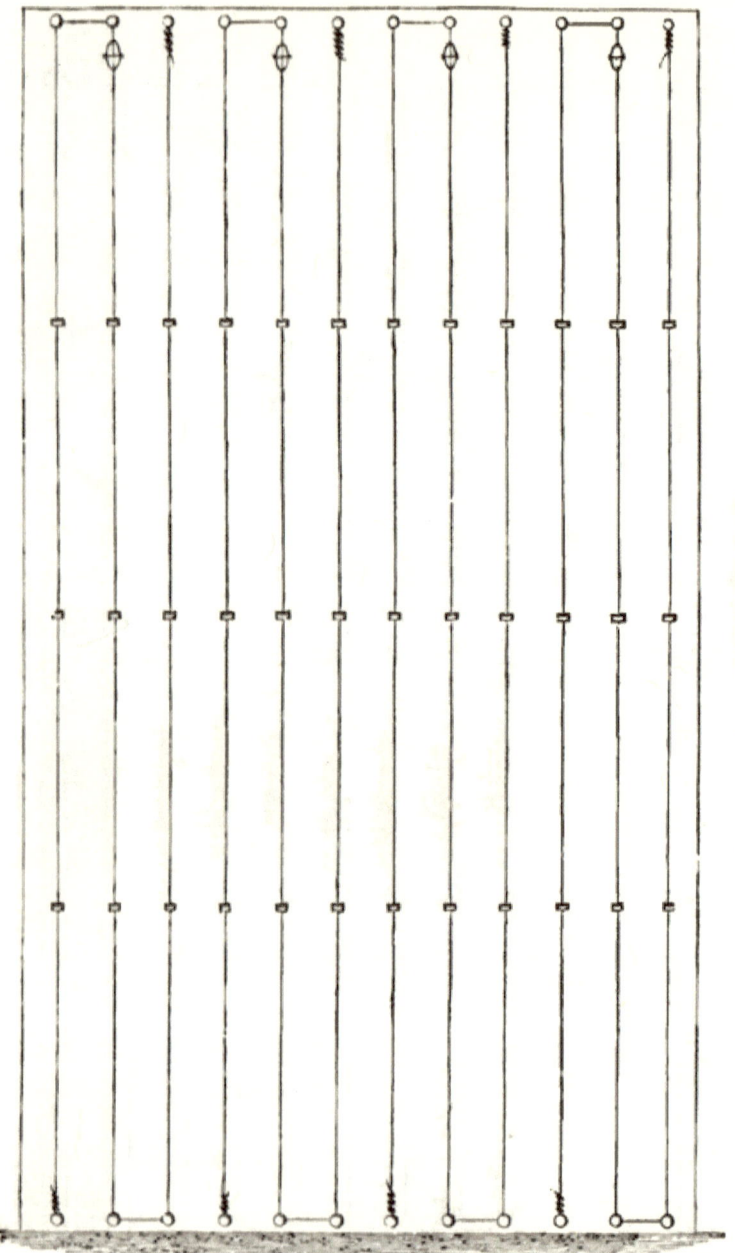

Fig. 108.—Iron Trellis for Vertical Cordon.

Standard Pear Trees.

Standard trees are not adapted for gardens; their proper place is the orchard, where they should stand ten or fifteen yards apart. Standards being too frequently neglected, left entirely to themselves in fact, often grown on one side, the branches run into confusion, and the middle of the tree becomes so thick that the light cannot penetrate; and this part of the tree remains completely barren.

To prevent these inconveniences, the tree must be trained so that the principal branches at the summit ray out regularly from the trunk, first in a horizontal direction, then a little depressed, then vertically. The head of the tree being kept open, the branches will grow somewhat in the form of a goblet. There will be as great an extent of branches if trained in this regular manner, and the light being able to penetrate the interior surface, the produce will be doubled.

In order to obtain standard trees of the form just described, the following course must be pursued:—

We suppose the trees to have been grafted, standard high, a year after plantation in the orchard. The first year after grafting we only allow two, three, or four shoots to grow, according to the degree of vigour of vegetation; the shoots must be left to grow at equal distances round the stem. The superfluous shoots must be pinched off when they have grown about four inches. The equal growth of the reserved shoots must be carefully maintained—one must not be allowed to become more vigorous than another: this must be regulated by means of pinching.

In the following spring, if three shoots have been preserved, the tree will resemble fig. 109. The

Fig. 109.—Tall Standards, First Year of Grafting.   Fig. 110.—Tall Standards, Second Year of Grafting.

branches must then be cut back at A, about eight inches from their spring, above two buds growing from each side the branch, which alone should develop vigorously during the following summer. All other shoots must be pinched off when three inches grown, and an equal vigour must be maintained between the six reserved shoots; at the third spring the young tree will resemble figs. 110 and 111. The shoots must be again cut back about sixteen inches from their spring, above two buds as before. The other operations must be repeated during the summer.

By the fourth year the head of the tree is composed of twelve principal branches growing at regular distances from and around the tree. These operations are

Fig. 111.—Plan of Fig. 110.

sufficient to impart a good form to the tree, and little more is necessary than to maintain an equal vigour in the growth of the branches, and to remove, about the end of May, the vigorous shoots which spring from the base and interior face of the principal branches. These shoots weaken the branches, and disarrange and confuse the orderly growth of the tree.

The formation and treatment of fruit-branches must, in the case of standard pear trees, be left to nature.

If, instead of grafting after one year's planting in the orchard, it is more convenient to plant trees already grafted, it will be well to choose such as have been grafted only one or two years, and furnished with at least two well-placed principal branches, to serve as a foundation for establishing the wood of the tree. After planting, the branches must be cut back only one-third of their length. The first pruning must not be applied until the following year, when the useless branches must be suppressed, and those reserved cut back to make them fork, as before described.

THE APPLE.

Soil.—The apple will grow in more humid soil than is suitable for the pear, but it prospers most in soils of medium consistency, rather sandy, but moderately cool.

Choice of Trees.—The directions already given for pears apply equally to apple trees.

Grafting.—The apple is grafted upon apple stocks grown from *pépins*, upon *Doucin* stocks grown from layers, and upon paradise stocks also grown from layers.

The most vigorous stock of the three is the apple,* which is only employed when high-stemmed *standard* trees are desired. The *Doucin* stock is less vigorous, and is chosen for pyramid, espalier, and vase-formed trees. The paradise stock is used for dwarf trees, trained in the form of small vases or bushes, the fruit of which is abundant and of excellent quality, and appears at the third year. Unfortunately these dwarf

---

* Apple stocks, also called free stocks, are used for standards for orchard planting. They are raised in numbers from the residuum in cider-making. Apples grafted upon these stocks are much longer in coming to bearing. The best stock for apples is the crab; the trees are hardier, and the fruit better. The Doucin or French stock is not equal to the paradise for dwarf trees. The paradise stock is chiefly used for trees to be grown in pots, for miniature orchards, and for ornamental trees in the flower-border.

trees are of much shorter duration than those grafted upon the other stocks.

The directions given for choice of grafts and modes of grafting pears apply also to apples.

Varieties.—Although less considerable than that of pears, the number of varieties of apples for table use is sufficiently extended. We have counted as many as five hundred varieties, amongst which we consider the following to be the best, affording a supply for the table for each month in the year.

Names of Varieties and Synonymes.	Name best known by in England.	Time of Maturity.
Calville rouge d'été .. *Tasse-pomme rouge.*	*Tasse Pomme*	August and Sept.
Borowiski	*Borovitski*.......... Dessert. Good.	End of August.
Monstrueux pépin	*Gloria Mundi* ...... Kitchen. Inferior.	Sept. and Oct.
Louis XVIII. ...... *Belle Dubois.*	....	October.
Rhode Island Gloria Mundi. Pater Noster.	*Rhode Island* ...... Greening culinary.	November to March.
Reinette Blanche .... *Reinette d'Espagne. Reinette tendre.*	*Cobbett's Fall Pippin* Very large & good.	Oct. and Nov.
Quatre goûts côtelée.. *Pomme violette. Calville rouge d'août. Pomme grelot.*	Red Calville	Oct. and Nov.
Calville de St. Sauveur.	....	November.
Belle Joséphine ...... *Ménagère.*	Hans Mutterchen ..	November.
Brabant belle-fleur ..	Red Belle-Fleur	Nov. and Dec.
Reinette d'Angleterre *Pomme d'or.*	....	Nov., Dec.
Citron .... *Reinette dorée.*	Citron	December.
Golden Pippin ...... *Rousse jaune tardive.*	Very fine dessert..	Nov., March.
Cornish gillyflower ..	Cornish Gillyflower Very excellent.	Dec. to Feb.
Pigeon d'hiver ...... *Gros pigeon.*	Arabian Apple......	Dec. to Feb.

THE APPLE.

Names of Varieties and Synonymes.	Name best known by in England.	Time of Maturity.
Gravenstein	*Gravenstein*	Dec. to Feb.
Reine des reinettes ..	Royal Reinette	Dec. to Feb.
Reinette gr. du Canada	*Reinette gr. du Canada* Very good.	January to March.
Reinette du Canada bl.	St. Helena Russet ..	January to March.
Royale d'Angleterre ..	Herefordshire Pearmain.	January to March.
Grosse reinette d'Angleterre.		
Calville blanc d'hiver *Bonnet carré.*	*Calville Blanc* ...... Bedfordshire Foundling.	January to March.
	Kitchen. Very good	January to March.
Api gros	....	January to March.
Reinette de Hollande	*Holland Pippin* .... Kitchen.	January, March.
Reinette du Vigan....	....	February to May.
Reinette fr. à côtes ..	....	February to May.
Reinette franche ordin.	....	February to May until August.
Reinette gr. h. bonté *Reinette de Rouen.*	....	February to May until July.
Reinette de Caux	Reinette Carse......	February to May.

The Editor recommends the following list of dessert and kitchen apples, adapted for general planting throughout England. In Scotland, the later varieties would require to be planted against a wall.

SUMMER APPLES.

DESSERT—
    Astrachan.
    Borovitski.
    Devonshire Quarendon.
    Jenneting.
    Margaret.

KITCHEN —
    Keswick Codling.
    Outhenden or Hawthornden.
    Lord Suffield.
    Manks Codling.
    Springrove Codling.

AUTUMN APPLES.

DESSERT—
- Blenheim Orange.
- Coe's Golden Drop.
- Cox's Orange Pippin.
- Early Nonpareil.
- Fearn's Pippin.
- Golden Pippin.
- Golden Reinette.
- Margil.
- Ribston Pippin.
- Stamford Pippin.
- Reinette Van Mons.
- Sykehouse Russet.

KITCHEN—
- Bedfordshire Foundling.
- Cellini.
- Cox's Pomona.
- Gloria Mundi.
- Golden Noble.
- Greenup's Pippin.
- Harvey's Apple.
- Lemon Pippin.
- Nelson's Codling.
- Tower of Glammis.
- Winter Quoining.
- Wormsley Pippin.

WINTER APPLES.

DESSERT—
- Braddick's Nonpareil.
- Claygate Pearmain.
- Court-pendu plat.
- Cornish Gillyflower.
- Cockle Pippin.
- Dutch Mignonne.
- Golden Harvey.
- Golden Russet.
- Keddleston Pippin.
- Mannington's Pearmain.
- Old Nonpareil.
- Pearson's Plate Apple.
- Ross Nonpareil.
- Spring Ribston Pippin.
- Sturmer Pippin.
- Wykin Pippin.

KITCHEN—
- Alfriston.
- Beauty of Kent.
- Bess Pool.
- Dumelous's Seedling.
- Gooseberry Apple.
- Norfolk Biffin.
- Northern Greening.
- Reinette Blanche d'Espagne.
- Rhode Island Greening.
- Royal Pearmain.
- Royal Russet.
- Striped Biffin.
- Winter Majeting.
- French Crab.
- Kirk's Lord Nelson.

The *Canadian Golden Reinette* requires an east or west wall to ripen in the midland counties. The *Api petit* ripens very well as a dwarf bush, but is inferior except for its great beauty. It is of a very red colour and a great bearer.

Training and Pruning.

The apple may be cultivated like the pear, either as standard or espalier; the form of standard, however, is better suited for the apple than that of espalier. The apple will not endure exposure to great heat so well as the pear; it requires a cool and rather humid air. Some varieties, such as the Canadian and golden reinette, white calville, api, winter pigeon, &c., &c., endure heat better, and may be planted as espaliers, but facing the west.

Training as Espaliers.—Any of the forms described for pear trees may be adopted for apples, and the trees should be planted in the same manner. The treatment of the fruit-branches is also exactly the same as described for pears.

Standard Trees.—The pyramidal form might, perhaps, be applied to apple trees, but it is not so well adapted for them as for pears. The *vase* or *goblet* form, described page 82, is more suitable, or the *double espalier*, page 89. The directions for forming them are the same.

The form recommended is that of the small goblet or bush, more or less regular, for trees grafted on paradise stocks. These small trees are extremely fertile, but, unfortunately, are not very long-lived. They are planted a number together about four feet apart. The cultivation of these small trees has been brought to great perfection by training them in single horizontal lines, which we shall now proceed to describe (fig. 112).

Select trees grafted on paradise stocks of one year if the soil is of good quality, or grafted on French stocks if the soil is parched and dry. Plant in a straight line five feet apart if paradise, and eight feet if on French [crab]-stocks. Cut off one-third of the length of the stem, and then leave the plants to themselves during the summer. The following year, at the winter pruning, place a galvanised iron wire (A), securely fastened at each extremity, along the line of the trees; this must be made tight by means of the tightener B, and supported every twelve feet by a small post (C) about eighteen inches above the ground. The wire being properly fixed, bend the branches of the trees in a horizontal direction, and fasten them to the wire. During the following year remove all the shoots which spring from the upright portion of the stem; they absorb too much of the sap, to the deprivation of the horizontal branches. Apply to the other shoots the treatment described for pears, in order to transform them into fruit-branches. Leave the terminal shoot to grow as much as it will during the summer. At the winter pruning treat the fruit-

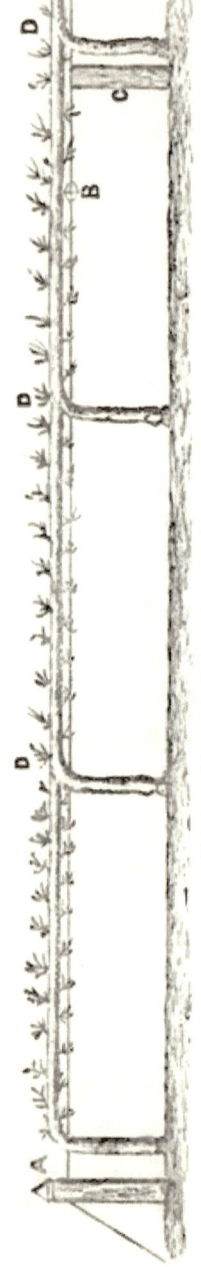

Fig. 112.—Low Apple Espalier for Border Edging.

branches as directed for pears; leave the new extension of the terminal branch entire, and fasten it to the wire.

The same treatment must be continued until each stem has grown sufficiently long to reach the next stem, and beyond it, to the extent of about twenty

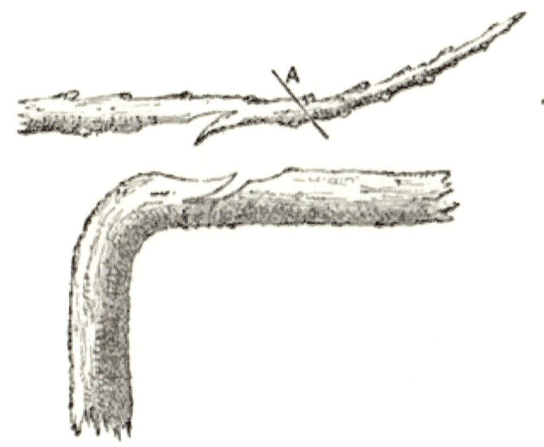

Fig. 113.—Mode of Inarching.

inches; it must then be grafted to it by inarching, at the point D, fig. 112. The proper time for this is the month of March; and the mode of operating is shown at figure 113. The following year the stems will be completely united, and the extremity must be cut off at A, figure 113. The result will be that all the superabundant sap of one tree will flow to the next one, and so along the whole line, all the stems being maintained in the same degree of vigour.

Such lines of small apple trees are easily formed, and come to fruit-bearing in the second year after planting. They may be placed in the edges of borders where espaliers have grown, and occupy places that could not be more usefully employed. Their diminutive height will form no impediment to the other trees upon the border.

THE PEACH.

Soil.—The peach requires a deep open soil, of medium consistency, containing a certain proportion of calcareous matter, and, which is very important, the soil must be free from all superabundant moisture. Soils which do not possess these qualities must have them imparted by other soils being added, and thoroughly mixed by deep digging.

The peach is almost always cultivated as a wall tree; the aspect may be east, south, or south-west, but it prefers the south-east.

Choice of Trees. — Peaches are generally planted already grafted. They should be grafts of only one year, healthy, vigorous, and with well-formed buds at the base. The kind of stock in which the peach is budded has much to do with the success of the tree.

The stocks most suited for the peach and nectarine, for out-door planting in England, are the muscle plum, St. Julian plum, and pear plum—the two latter for the more delicate kinds, and for pot culture, but the muscle for general planting.

For deep dry soils the almond forms the best stock, for more humid soils the plum, the roots of which are less liable to strike deeply into the soil than the almond; unfortunately these trees are less vigorous, and shorter lived.*

Grafting.—These stocks may be grafted by the shield graft, double or single, the almond at the beginning of September, the plum in July.

Varieties.—There are fifty varieties of the peach at the present time, but many of them are only suitable for the climate of the south of France, and others are only of inferior quality. The following is a list of the best sorts arriving at maturity at different periods of the fruit season :—

Names of Varieties and Synonymes.	Observation.	Time of Maturity.
Desse hâtive		End of July.
Grosse mignon. hâtive		Beginning of August.
Pourprée hâtive		Middle of August.
Grosse mignon. tardive		End of August.
Belle Bausse		End of August.
Reine des vergers		Beginning of Sept.
Madeleine r. courson		Middle of September.
Lisse gros. violette hât. *Violette de courson.*		Middle of September.
Brugnon de Stanwick		Middle of September.
Admirable jaune		End of September.
Admirable *Belle de Vitry.*		End of September.
Bourdine de Narbonne *Grosse royale.*		End of September.
Chevreuse tardive *Bonouvrier.*		End of September.
Desse tardive	Good late Peach	Beginning of October.

* The almond would do little or no good in this country on its *own roots;* it is always propagated by budding on the Brussels plum stock, or on the Brampton plum stock. The Muscle plum stock is the principal stock for peaches and nectarines suitable for English planting. The peach is certainly the finest fruit produced in the open air in England, and a practical knowledge of the best manner of bringing it to perfection is most desirable; there cannot be too much care bestowed in the propagation, culture, and management of the many fine varieties we possess.—Ed.

THE PEACH. 137

The following is a list of some of the best kinds of peaches and nectarines now in cultivation in England :—

Names of Peaches.	Observations.	Time of Ripening.
Small Mignonne	Well adapted for culture.	August.
Early Grosse Mignonne	....	August.
Abec	Ripens well	August.
Crawford's Early	Very large and good	September.
Grosse Mignonne	....	September.
Royal George	....	September.
Noblesse	....	September.
Bellegarde	....	September.
Barrington	....	September.
Walburton's Admirable	....	October.
Catharine	....	October.
Violette hâtive	....	September.

Names of Nectarines.	Time of Ripening.
Elruge	August and September.
Hardwicke Seedling	August.
Rivers's Orange	September.
Roman	September.
Violette hâtive	August and September.
White	September.

TRAINING AND PRUNING.

The peach may be trained to various forms; but we shall only describe two, the Verrier palmette, and the simple oblique cordon, because these forms are more easily obtained, and may be best accommodated to local requirements.

TRAINING THE PEACH IN THE PALMETTE FORM.

Formation of the Wood.—The treatment is in all respects the same as described for the palmette form of the pear (fig. 84, p. 96), the only difference being that in the pear the side branches are trained twelve inches apart, while the side branches of the peach must be from 20 to 24 inches apart, in order to allow sufficient space for nailing up the lateral shoots during the summer. Besides this, all the branches of the peach, both the main stem and side branches, are furnished on each of their sides with fruit-branches, about four inches apart.

The trees must be planted at such distances as to cover a medium surface of about 20 square yards of the wall. Thus, for a wall three and a half feet high, it will be necessary to plant about six yards apart.

First Pruning.—Pears, and other trees of the same species, ought not to receive their first pruning until they have struck root, that is, about a year after planting; the peach, however, must be pruned the same year that it is planted, otherwise the buds at the base of the stem which should be developed into shoots would be completely withered by the following year.

The first pruning has for its object to develop towards the base of the tree the first two side branches, and to obtain a new extension of the stem. To effect this, select two lateral buds, B (fig. 114), situate about twelve inches from the ground. Also the bud A above and in front; it is immediately above this latter bud, at the point D, that the stem must be cut. The buds B are intended to form the first side branches, and the bud A the extension of the stem.

The buds selected for the extension of the stem and for the side branches should be as much in front as

THE PEACH.   139

possible; by this means the slight deformity at each point, from which the new extension springs, is less apparent, and the place of the cut is better shaded from the sun, and more readily heals.

During the following summer the development of the three buds must be watched; if other branches

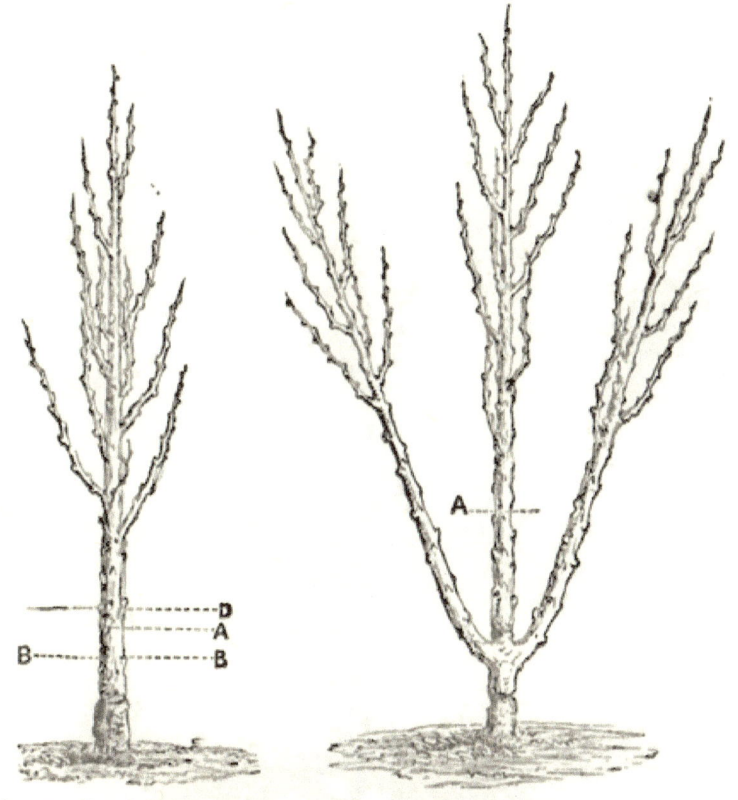

Fig. 114.—Peach, Palmette Form, First Year.   Fig. 115.—Peach, Second Year.

grow upon them, they must be pinched when they have grown to six inches, and be suppressed entirely when the reserved shoots have attained a length of 16 inches. The side branches must be maintained at an equal degree of vigour by the means described in the chapter on the PRINCIPLES OF PRUNING.

Second Pruning.—Figure 115 shows the result of the operations of the preceding year. At the second pruning, about the third part of the length of each side branch must be suppressed, selecting for the place a front bud suitable for the new extension. The main

Fig. 116.—Peach, Third Year.

stem is cut at A, about twelve inches above the spring of the side branches, and immediately above a front bud. The stem might be cut higher up, with a view to another set of side branches during the summer; but it is more prudent to allow an interval of two years between the formation of the first and second sets of side branches. By this means the growth of the lower branches will be best promoted, they having always a tendency to be less vigorous than the higher ones.

During the following summer the terminal buds must receive such attention as is necessary to keep them in the same degree of vigour. The other buds must be treated as described on p. 146, in order to transform them into fruit-branches.

Third Pruning.—At the third spring the young peach tree will resemble fig. 116. The main stem must now be cut 24 inches above the spring of the side branches at A, above the two side buds B and C, intended to develop two new side branches, and a bud in front to prolong the main stem. The side branches must be cut back about one-third of their new growth at D, in order to make them develop all their remaining buds.

In pruning the side branches, it is of importance to give the same length to each of the parallel branches, in order to maintain an equilibrium between both sides of the tree. If one side branch has become stronger than the corresponding one, it must be cut rather shorter. The fruit-branches growing towards the lower part of the tree must be treated as described further on (p. 152).

During the whole of the summer the same treatment must be applied to the principal shoots as in the year preceding.

Fourth Pruning.—The result of the operations of

the previous year are a new set of side branches, as shown by fig. 117. These must be cut back about a

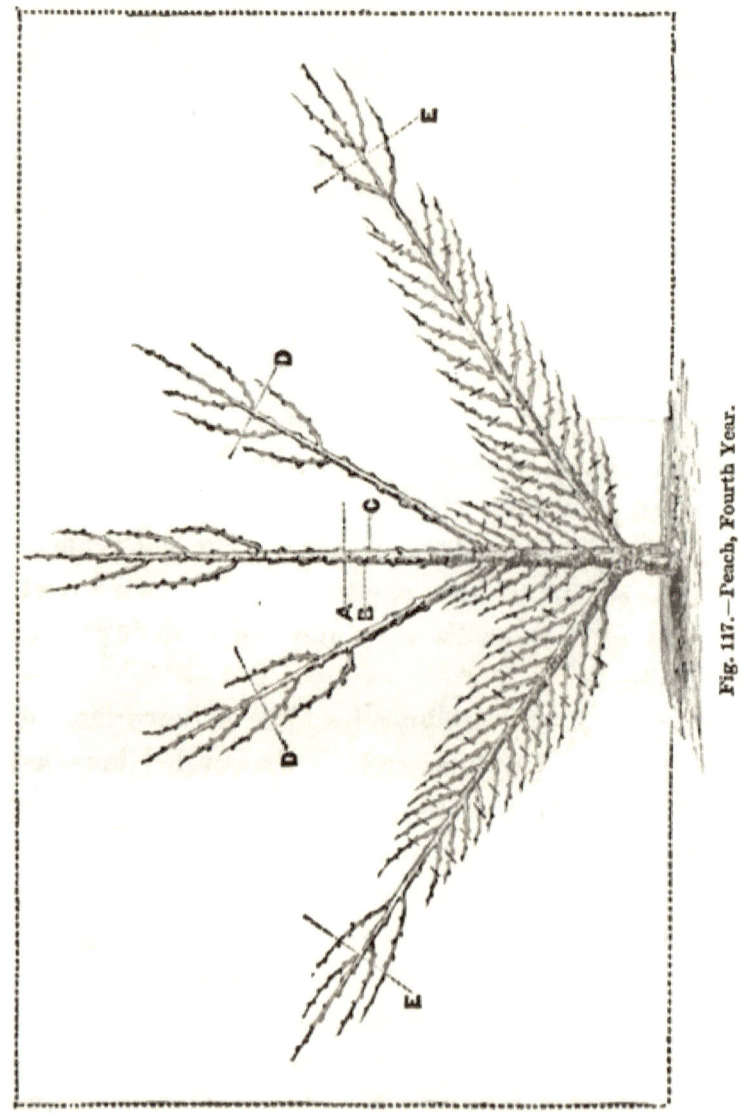

Fig. 117.—Peach, Fourth Year.

third of their length at D; and the new growth of the lower side branches must also be cut back one-third

THE PEACH.

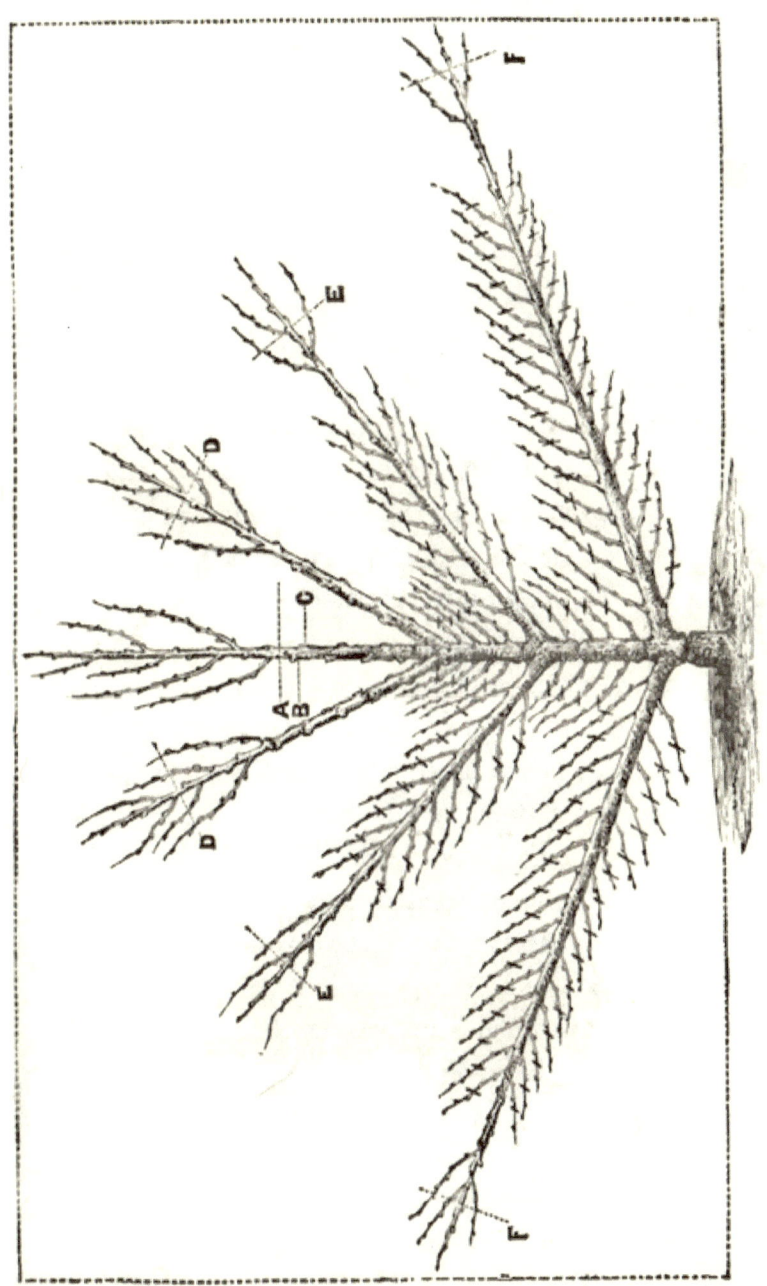

Fig. 118.—Peach, Fifth Year.

at E. The new extension of the main stem must be cut at about 24 inches above the spring of the upper side branches at A, so as to obtain, by means of the buds B and C, a new set of side branches. A new series of side branches may from this period be obtained every year; for the lower side branches have now acquired sufficient strength to draw to themselves all the sap which they require for their continued growth. The summer treatment is the same as before.

Fifth Pruning.—A third set of side branches have grown during the preceding summer (fig. 118). The new extensions of the side branches must be cut back as in previous years. The main stem must be cut at A, and a new set of side branches obtained from the buds C and D. The remaining treatment as before.

The operations we have described must be continued from year to year, in order to obtain new side branches and regular extensions of them, until they acquire the desired length, when their ends must be directed upwards, as shown (figs. 84 and 89), until each of them attains the summit of the wall. Towards the tenth or twelfth year the trees treated in the manner we have described will assume the aspect of fig. 84, with this difference, that the fruit-branches are distributed on each side of the wood-branches, as shown at fig. 118.

Nailing up.—If it is not wished to fasten up the branches with list, as described at p. 104, it may be done by forming a trellis. It is necessary to fasten up not only the wood-branches of peach trees, but, as we shall see further on, the fruit-branches after the winter pruning and the lateral shoots during the summer. As the points where fastenings are required are very numerous, the trellis bars must be rather close together, and for this reason it will be expensive if made

Fig. 119.—Trellis of Wood and Iron for Peach.

altogether of wood. A wire trellis will be preferable (fig. 94, p. 106), having spaces of 32 inches between each of the horizontal lines. If there is already upon the wall a wooden trellis with wide squares, it may be used, dividing the squares with lines of iron wire (fig. 119).

Obtaining and Replacing Fruit-Branches on the Peach Tree.

There is an important difference between *pépin* fruits and stone fruits. In the former the fruit-spurs require about three years for their growth, but when formed they last for an indefinite period with proper attention. In stone fruits, and especially in the peach, the fruit-spurs expand their blossoms in the spring following their birth, but they produce no more blossoms. Those which appear the following year are from new fruit-buds which have grown during the preceding summer upon the primitive branch. It follows, therefore, that the first thing to be attended to in the management of peaches is to obtain the fruit-spurs, and to replace them year by year.

We have already said that the fruit-spurs must be made to grow regularly from each side of the branches, about four inches apart, in such a manner that each branch shall be in the form of a fish's skeleton. This result is obtained as follows :—

First Year.—We take, for example, any extension of a branch developed during the preceding summer (fig. 120). At the winter pruning a part of the new extension is cut back, in order to make it develop all the buds that it carries. Unless this operation were performed, a number of buds at the base of the tree would remain inactive, and there would be a void upon the branch very difficult to fill up, for the buds that had not been developed the first year would be useless the next. Towards the middle of May the extension presents the appearance of fig. 121; all the buds have expanded into shoots. When the shoots have attained a length of about three inches, the useless buds must be removed, including such as grow in front (A, fig. 121) or at the back of the branches. There is no exception to this, unless some of the side buds are too wide apart, in which case a front or back bud may be left to supply the vacant place (C D, fig. 121). If there is any choice, the bud at the back of the branch is the better one to leave, as the irregularity will be less apparent.

The extensions of the branches generally present single wood-buds (A, fig. 120), but they are frequently double, and even treble, B. It is necessary to leave a single bud only at each point. If these double or treble

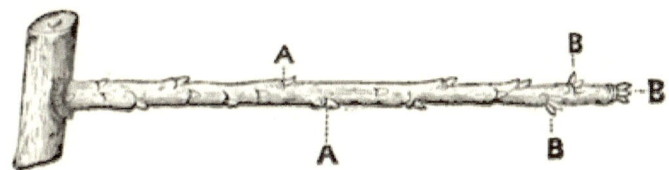

Fig. 120.—Peach, Wood Extension.

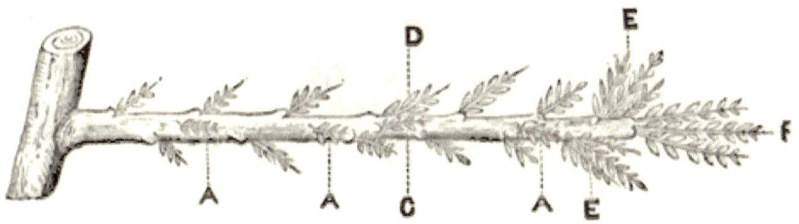

Fig. 121.—Peach, Wood Extension.

buds occupy the place of a fruit-branch, the weakest bud (E, fig. 121) must be retained, as in this case there is greater reason to fear an excess of vigour than of weakness. If it is desired to increase the length of the branch, the most vigorous bud, F, must be retained. The buds must not be torn off, but cut away with a sharp knife.

The buds that are retained must not be abandoned to their own growth, for many will become too vigorous, to the injury of the terminal bud, which ought to preserve the pre-eminence; and they will seldom produce fruit-blossoms. It is, therefore, necessary to restrain the vigour of the too luxuriant shoots, in order to preserve the general symmetry of the tree, and the proper direction of the branches.

The first of these objects is obtained by pinching.

Fig. 122.—Peach Bud Pinched.

The shoots that grow too vigorous upon the upper side of the lateral branches, and those which are too near the summit of the vertical branch, must be pinched

Fig. 123.—Gourmand Bud Pinched.

Fig. 124.—Result of Pinching No. 123.

back at A (fig. 122), when they are of the length of ten or twelve inches.

Buds that indicate by their vigour that they will

become *gourmand* branches (fig. 123) must be cut at A, above their two lower leaves. This will give rise to buds which will become shoots (B, fig. 124), and which can be treated as fruit-branches at the winter pruning.

The buds that are less vigorous should not be pinched, unless they have grown beyond sixteen inches.

The first pinching generally suffices to arrest the excessive growth of shoots intended to become fruit-branches; but they not unfrequently afterwards send

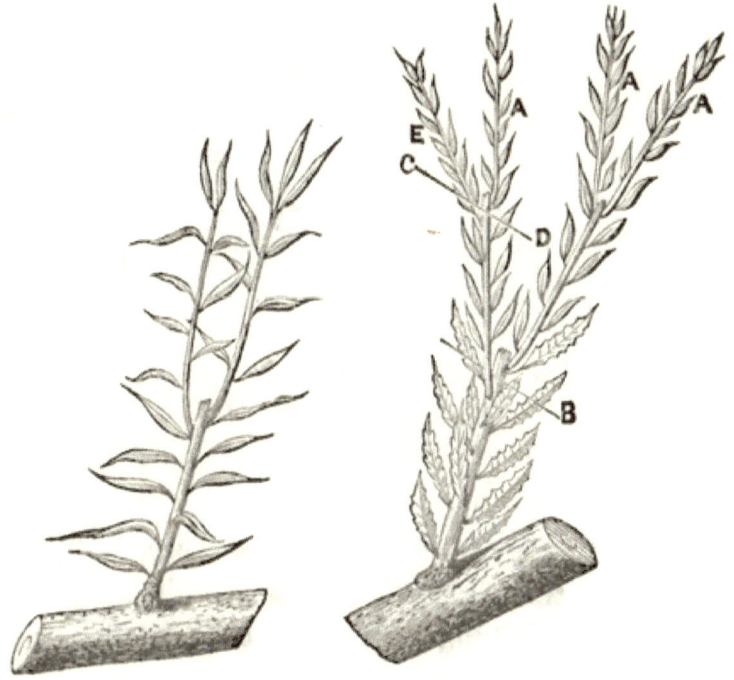

Fig. 125.—Pinching Irregular Buds.

Fig. 126.—Shoot bearing a succession of Irregular Shoots.

out towards the top one or two irregular branches (fig. 125), which must be pinched when grown about eight inches. It is seldom that pinching has to be resorted to a third time. Should this, however, occur, as A (fig. 126), the main shoot, must be cut off at B, then the shoot C D. The shoot E, which is reserved, is at the same time to be pinched. By this means confusion is avoided at the nailing up in summer.

When the *gourmand* shoot (fig. 127), which extends each branch, has attained a certain length, it will develop irregular shoots; these may be disbudded and pinched. This mode of operating, however, only gives rise to badly-formed fruit-branches for the winter

Fig. 127.—Extension bearing Irregular Shoots.

pruning. It will be preferable to proceed thus :—When they show their second pair of leaves (E, fig. 128), cut them with the nail below these last two leaves. Their vegetation is thus suspended, and there will be by the time of winter pruning a very short branch, much preferable to the result of the former process.

It will be better not to apply these operations upon the shoots further back than the point to which it may be expected that they will be cut back at the winter pruning.

We have said that it is most important to train these

shoots in their proper direction, which must be done at the time of nailing up in summer. During the summer, the shoots which form the extensions of the wood-branches must be nailed up as soon as they are twelve inches long. The most vigorous of the lateral shoots must be nailed up when they are grown about ten inches, and the feebler ones when they are fourteen inches. The shoot must be fastened in such a direction

Fig. 128.—Irregular Bud, Pinched early.

as to form an acute angle with the branch upon which it grows. Care must be observed not to enclose the leaves in the list, and not to make the shoots cross each other. Nails and list are used for fastening, if the wall is suitable for nailing; a green rush is used if the trees are supported by a trellis.* The nailing

* A green willow twig is generally used for fastening to a trellis, or prepared wire.

up, or fastening, must be done progressively (not all at once, as is too often done), in order to keep the different shoots of an equal vigour.

Second Year.—The operations of the summer are intended to transform the shoots into branches, like those we shall now describe.

The shoots growing below the side branches, near their spring, often remain nothing more than very short branches, bearing scarcely anything except blossoms, and a wood-bud at the end (fig. 129). These minute fruit-branches do not require pruning; the fruit which grows upon them is the best which the tree produces.

Some other shoots, rather longer, but growing also in unfavourable situations, have become branches of from six to eight inches in length, covered with flower-buds, except towards the base, where may be seen two or three wood-buds (fig. 130); these are the fruit-branches proper. They must be cut back, in order to obtain a new well-placed fruit-bearing branch for the next year; but some of the flower-buds must be reserved, to insure fructification.

In order to prove, by an example, the absolute necessity of cutting back these fruit-branches every year, we will suppose that the branch *a* (fig. 130) has been left unpruned: it will bear fruit the first year, and then put forth at the top one or two shoots, which will become branches the following spring, and upon which only will appear flower-buds, for, as we have observed, the fruit-branches of the peach bear only once. This ramification will present in the following spring the appearance of fig. 131. If left to itself again, the same results will follow; and it will be seen, that if each of the side branches were left to grow in this manner, the branches would become so indefinitely prolonged, that it would not be possible for the sap to afford sufficient nutriment for them, and many would wither, especially

Fig. 129.—Small Fruit-Branch in Blossom.

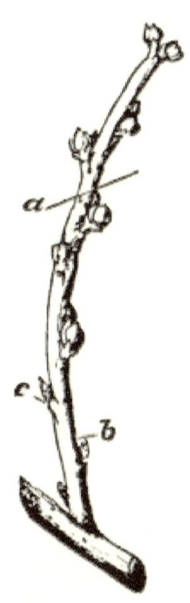

Fig. 130.—Ordinary Fruit-Branch of Peach.

towards the base; and the vacant places would seriously detract from the appearance and regularity of the tree. It is in this way that unpruned, or badly pruned, peaches are often destroyed.

Under these circumstances, we see how the branch *a* (fig. 130) ought to be pruned, retaining at the same time a sufficient number of flowers to determine the development of the wood-buds *b* and *c*. This double result will be attained by cutting at *a*, about three or four inches from the spring of the branch.

The flower-buds B (fig. 131) are almost always ac-

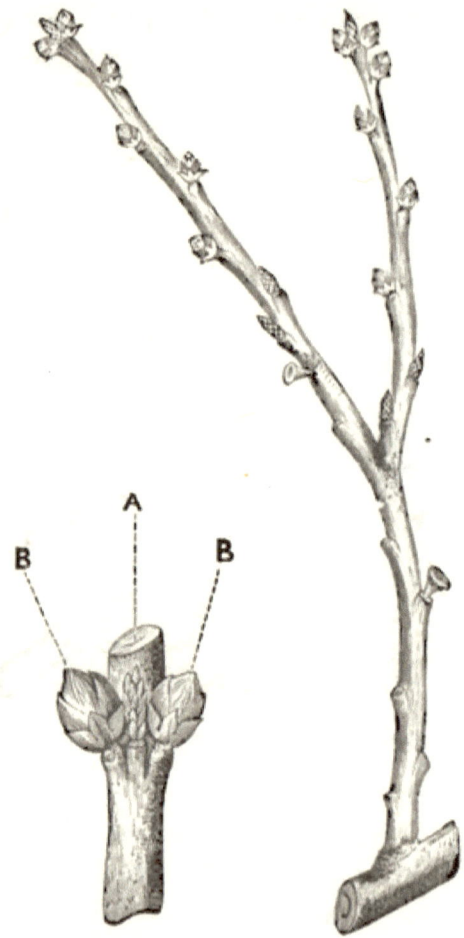

Fig. 131.—Wood-Bud and Fruit-Bud of Peach.   Fig. 132.—Neglected Fruit-Branch.

companied by a wood-bud A, and also sometimes by small branches without any wood-buds, unless there be one or two minute ones towards the base (fig. 133). It might be thought, that the blossoms upon these small branches, unaccompanied by a wood-bud, must prove

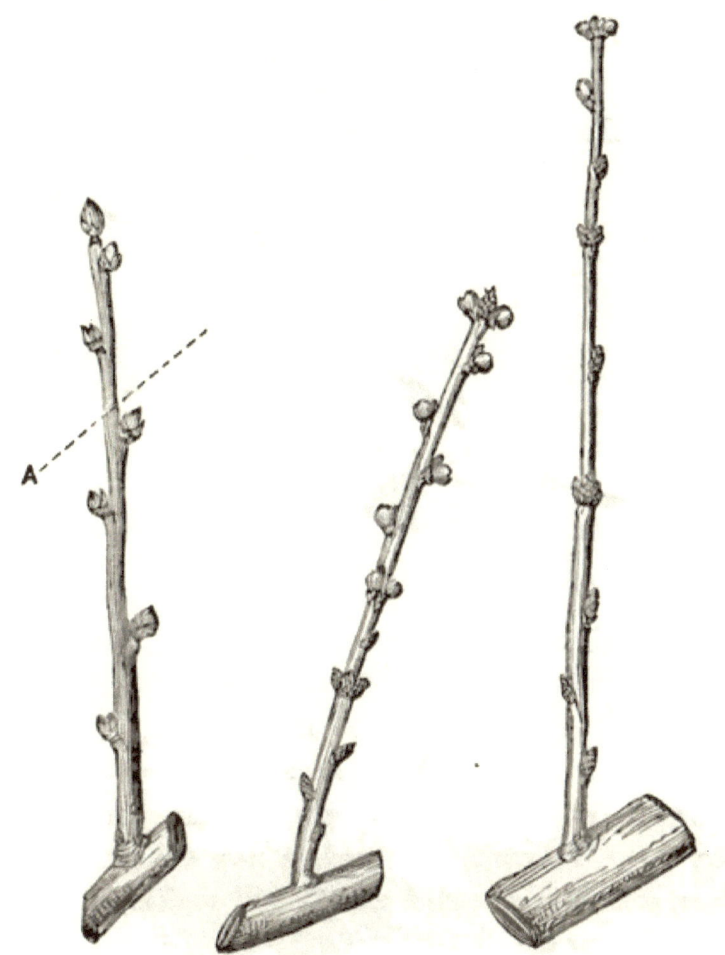

Fig. 133.—Small Irregular Fruit-Branch.   Fig. 134.—Fruit-Branch, Mixed Buds, First Pruning.   Fig. 135.—Wood-Branch, First Pruning.

sterile, and ought to be cut off at pruning, as though of no value; quite the contrary, however; experience proves that these blossoms produce the finest fruit.

These branches, therefore, must be preserved, and pruned at A.

Some of the buds more favourably situated produce more vigorous branches, and bear only wood-buds (fig. 134) for three or four inches from their base upwards. These must be cut above the second bud, in order to produce the result before described.

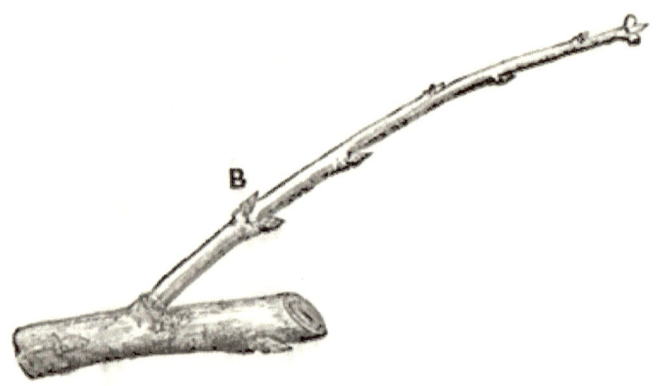

Fig. 136.—Irregular Branch, without Buds at Base.

If the shoots are still more vigorous than those which produce both wood and fruit buds, they will resemble fig. 135, and will produce wood-buds only, with, perhaps, one flower or so at the top. These must be cut above the second wood-bud from the base. If not cut low down, they will not produce new shoots at their base, and their extended growth will weaken, and perhaps destroy them.

In speaking of *gourmand* shoots, we referred to irregular shoots. If the *gourmand* shoots have been pinched off at too great length, they give place the following winter to irregular branches (fig. 136). These branches differ in structure from any that we have at present described. They are, in fact, almost without buds for three or four inches from their base; this is a very troublesome disposition, for whatever we may do, in the process of obtaining buds, the branch will grow to too great a length. Sometimes, however, these branches present two buds, as shown at A (fig. 137).

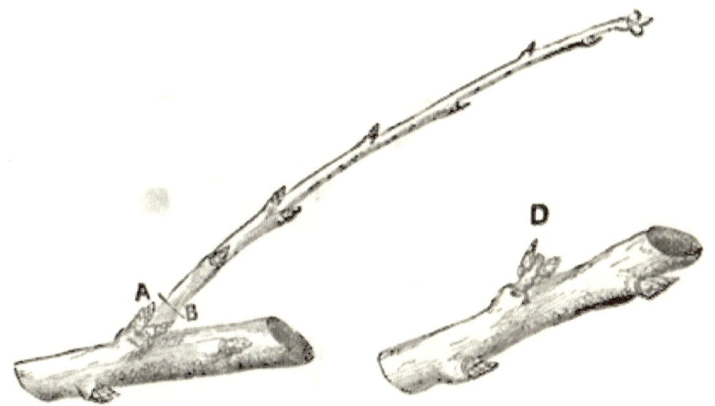

Fig. 137.—Irregular Branch, with Wood-Buds at Base.

Fig. 138.—Irregular Branch, Pinched too early.

These branches are cut at B, above the wood-bud nearest to the base. This close pruning, repeated during several years, often gives rise to new wood-buds at the point of junction with the main branch. But if the irregular shoots have been pinched much shorter than we have described, they will produce small branches (D, fig. 138), which need not be cut at all.

The different branches which we have now described are the only ones that should be met with on a carefully pruned tree. Unfortunately, however, the pinching is often not performed soon enough, and the branches will become *gourmand* shoots, which will take

158 FRUIT TREES.

the place of fruit-bearing branches (fig. 139). If these

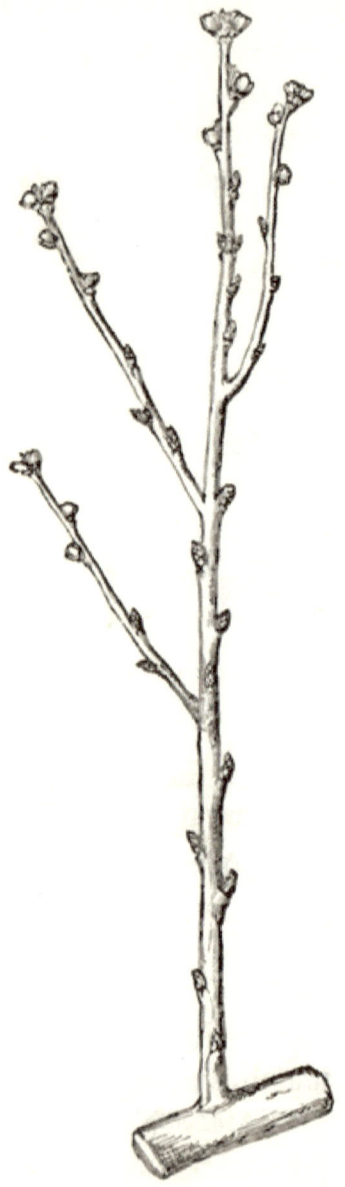

Fig. 139.—*Gourmand* Branch of Peach, First Pruning.

branches are cut above the two lower wood-buds, two

new branches will grow during the summer, which it will not be possible to prevent becoming too vigorous, the sap having taken so strong a current in that direction. The desired result will be better accomplished by *twisting* the branch about an inch above the base, and for four inches upwards; then cut between three and four inches above the twisting. One part of the sap will traverse the twisted portion, and will be lost above, and the lower buds, receiving only just sufficient for their development, will push out less vigorously, and give place the following year to two new branches, covered with fruit-blossoms. The primitive branch must now be cut immediately above the two

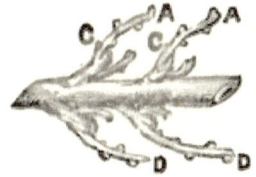

Fig. 140.—Fastening Fruit-Branches.

new branches, and all the twisted part will disappear. Instead of twisting, we may take off about half the thickness of the branch (on the wall side of the branch), about the same length as the twisted part mentioned above. The same effect will be produced.

When the fruit-branches have been pruned, and also the wood-branches, and the latter fixed to the wall, we proceed immediately to the winter nailing of the fruit-branches. The branches A (fig. 140), upon and above the oblique and horizontal branches, are bent so as to form a free curve. This form, very little deviating from its natural direction, tends to restrain the flow of sap towards the top of the branch, and to promote the development of the lower buds, which ought to produce the replacing branches.

The branches D, below the horizontal or oblique branches, must also be bent towards them as much as possible, to secure the same result.

Finally, the branches growing from the sides of the vertical branches ought to be nailed so as to form a right angle with the branch. If trained vertically, the

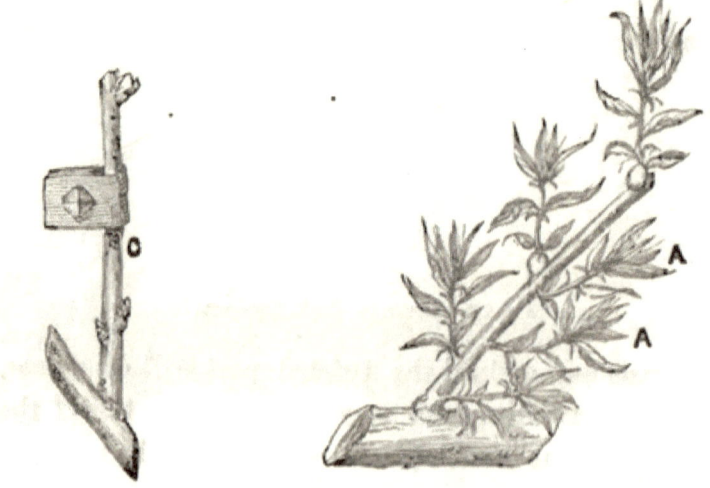

Fig. 141.—Fastening Fruit-Branches.   Fig. 142.—Disbudding Fruit-Branches.

action of the sap will favour the buds at the top, to the injury of those at the base.

Figure 141 shows how these branches are fixed by means of nail and list. Those grown upon a trellis may be tied with osier twigs. Some employ lead wire, which is quicker than the other mode, and neater, but rather more expensive.

During the ensuing summer the fruit-branches receive the series of operations we shall now describe. When the shoots have attained a length of two or three inches, they must be deprived of all their fruit-buds, except the two nearest the base, and each of those which accompany a fruit (fig. 142). The two buds A are suppressed to avoid confusion at the summer nailing up, and to preserve more vigour for the replacing shoots.

It may occur that some of the flower-buds reserved on certain branches at the winter pruning turn out

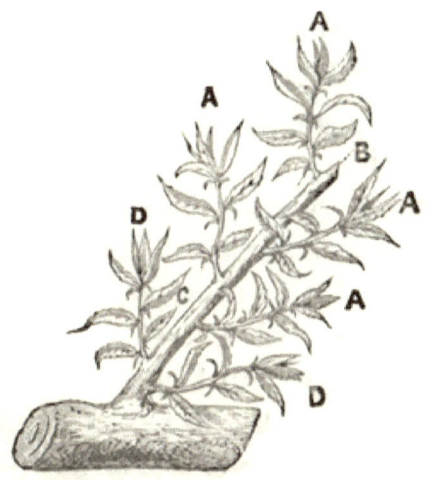

Fig. 143.—Green Pruning, First Year.

unproductive. These branches must be submitted to green pruning: thus, the branch B (fig. 143) being entirely fruitless, the shoots A, which would otherwise be reserved to nourish these fruits, are now useless.

The branch B is then to be cut at C, to preserve the two buds D, which will then take a more favourable form for developing the replacing shoots.

When the proper time arrives, apply successively pinching and the summer nailing; bearing in mind at all times that when the branches accompanied by the young fruits (fig. 142) have grown to six inches, they must be pinched in order to favour the development of replacing shoots at the base.

Notwithstanding every care to keep the two sides of the branches well furnished with fruit-branches, vacant spaces will occur owing to the destruction of buds or the decay of some of the fruit-branches. The best way of filling up these vacancies is grafting by herbaceous approach (page 5), which may be done at the time of summer nailing.

The remaining operations of the second summer relate to the fruit.

The superabundance of fruit is a more serious evil in regard to peach trees than to *pépin* fruits. When, therefore, the fruit is too abundant, it will be necessary to remove a certain number, leaving only half as many peaches as there are fruit-branches. This thinning must be made (when the fruit is about the size of a large walnut) from the lower side of the branches and lower part of the tree rather than from the upper.

When the peaches have nearly attained their full size, the leaves which shade them must be removed, in order to allow them to attain their natural bloom and colour; but the leaves must not be removed all at once, but at two different times on cloudy days, so

THE PEACH. 163

as to habituate the fruit gradually to the full influence of the sun. The leaves must not be carelessly torn off, but cut away so as to leave a small portion of each stem and leaf. If this is not attended to, the eye at the base of the leaf-stem will be lost, to the injury of next year's crop.

Third Year.—At the third year the winter pruning is practised as follows :—

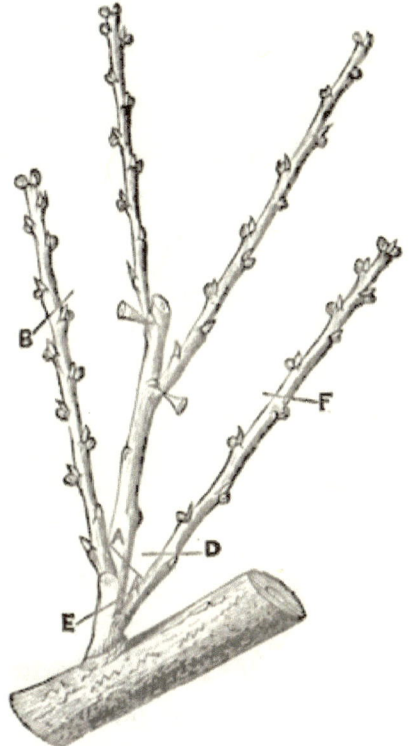

Fig. 144.—Fruit-Branch, Second Pruning.

The fruit-branches proper (fig. 130), which have borne during the preceding summer, have now grown to the form of figure 144. The fruit-branch is cut at A; the lower part E being intended to bear the succession of fruit-branches, is called the stock-branch. The branch B is selected for a new fruit-branch, and must be cut at B, reserving a number of the fruit-buds. The branch D, intended to replace branches that have borne once and are now useless, must be cut at D immediately above two wood-buds, the nearest to the base, which will furnish two new fruit-branches for the following year, and which must be cut as we shall describe. It follows that the stock-branch will bear every year two new fruit-branches, one furthest removed from the main branch, and which is cut long, while the other, the replacer nearest the base, is cut short above two of the lower buds. This is called crochet pruning.

Sometimes the branch B, the best situated for fruit-bearing, is without flower-buds. As it is too far from the main branch to be the replacer, we cut the primitive fruit-branch at E, and the branch D is cut at F above one or two flower-buds, to serve as replacers when wanted.

If there are no buds on any of these branches, we cut the primitive branch at E, and then the branch F at D.

All the other branches having received, during the preceding winter and summer, operations intended to give them the form we have described, the same mode of treatment must be applied to them.

It is important to remove every year, at the time of the winter pruning, the tail ends of the peach stems, which would otherwise injure the circulation of the sap. All dry little stems and twigs must now be removed.

The nailing or fastening up is the same as the first year. When summer comes the disbudding must be performed, only leaving, upon each fruit-bearing branch (fig. 145), the shoots which accompany one fruit; all others must be removed. Thus, in the figure 145, the three shoots C C and A must be removed; the branch B bears the two shoots which will replace the fruit-branches. If one of the two shoots on B does not develop itself properly, we must preserve the next one nearest to the base of the primitive fruit-branch.

If any of the flower-buds preserved upon the fruit-

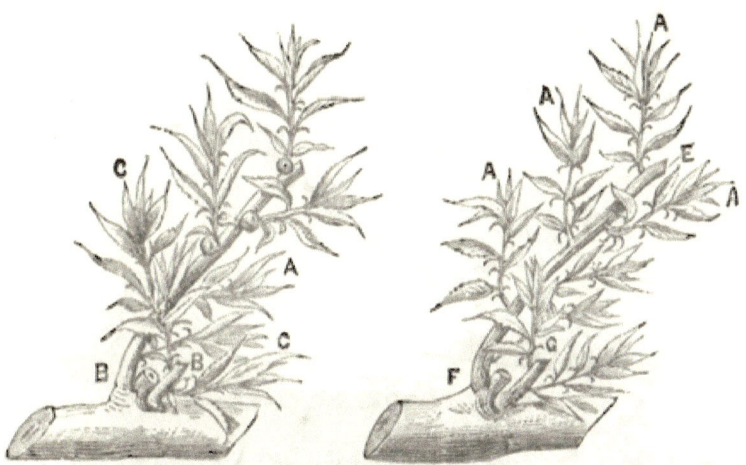

Fig. 145.—Disbudding, Second Year.   Fig. 146.—Spring Pruning, Second Year.

branch have not produced a fertile flower, the branch E must be cut at F (fig. 146), then the branch G insures the success of the replacing branches.

The disbudding and green pruning of fruit-branches

at the second year often involve the suppression of a third part of the shoot. This occasions much trouble, and the flow of gum is perhaps the result. It may be desirable, therefore, not to perform the operation all at

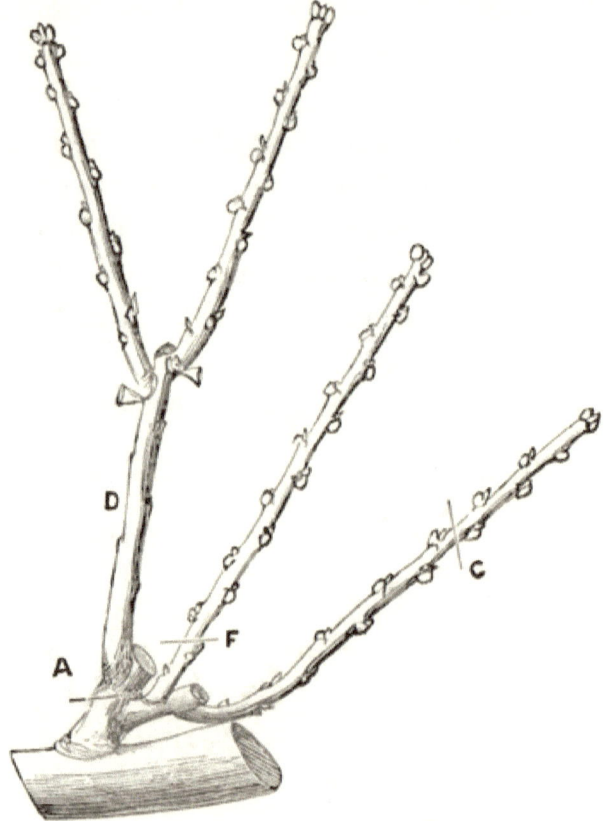

Fig. 147.—Fruit-Branch, Third Pruning.

one time, but at twice, first on the higher part of the tree, and eight or ten days later on the lower half. There is the additional advantage in this plan, that the sap drawn in such great abundance towards the lower part of the tree during these eight or ten days tends to augment the vigour of this part, always less favoured than the summit. The pinching, nailing up, removal of superfluous fruit and leaves, are all executed as during the preceding summer.

Fourth Year.—At the spring of the fourth year the branches that have been treated as fig. 144, and which have fructified during the summer, are now grown as

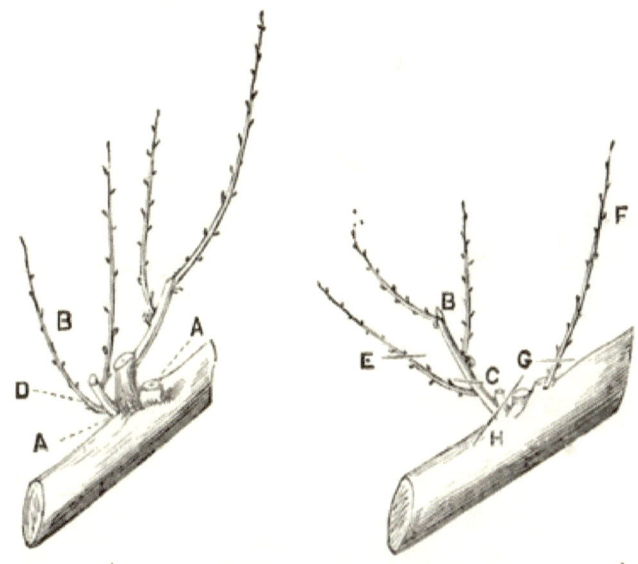

Fig. 148.—Renewal of Peach Branch.   Fig. 149.—Renewal of Peach Branch.

fig. 147. The stock-branch which bears the old fruit-branch D is cut at A; the branch F at F to form the replacer; the branch C at C for the new fruit-branch. This operation gives the same result at spring as before described. The same pruning must be applied from year to year. The other operations, both for summer and winter are the same as for the third year.

It frequently happens that the stock-branches of about three or four years' growth develop, towards their base, one or more wood-buds (A, fig. 148). We may take advantage of this to renovate the stock-branch that has become weak by successive prunings. For this purpose, instead of the crochet pruning, we preserve only the branch B, and cut it long for a fruit-branch. During the summer all the shoots that accompany a fruit must be preserved, besides the one which is nearest the base, and also one of the buds appearing at A. At the end of a year we have obtained the result represented by fig. 149. The primitive branch B is then cut at C, and the branch which it bears at its base at E: this last will be the fruit-branch. The branch F is cut at G above two wood-buds which furnish the replacers for the next year, at which time the stock-branch becomes useless and is cut off at H.

Training the Peach in Simple Oblique Cordon.

Ten or twelve years are generally required to perfect the wood of peach trees, trained as Verrier palmettes, or one of the other large forms now in use.

Now the average duration of a peach as an espalier is twenty years; it appears, therefore, that we employ the half of their existence to form their wood, and that half the surface of the wall remains unoccupied during five years.

Considering that the care and attention necessary to obtain these forms, even the least complex of them, are so complicated as to be beyond the time and means of ordinary gardeners, we now propose a method that will avoid these inconveniences altogether. This method (see fig. 150) is the same that we have re-

THE PEACH. 169

Fig. 120.—Oblique Cordon for Peaches.

commended and described for espalier pears, which we called the simple oblique cordon. We applied this form to peaches, in the year 1843, at the *Jardin des Plantes*, at Rouen. The operation is performed thus:—

We choose for planting, young trees of one year, from budding with only one stem (fig. 151); these are

Fig. 151.—Oblique Cordon, First Year.   Fig. 152.—Oblique Cordon, Second Year.

planted 30 inches apart, and inclined at first to an angle of 60 degrees. At the first pruning they must be cut at eight or twelve inches from their base, above a front wood-bud (A, fig. 151). If there are any irregular shoots below the cut, they must be entirely removed from before and behind; all the rest must be cut above the two wood-buds nearest their base.

During summer we promote as vigorously as possible the growth of the terminal shoot, and use the necessary means to transform the other shoots into fruit-branches.

The disbudding, green pruning, pinching, nailing up in summer, and the rest are all performed as directed for other forms of peach trees. In the following spring each of the young trees resembles the figure 152.

At the second pruning we take off about one-third of the entire length, cutting always above a front bud (A, fig. 152). The fruit-branches must be pruned and nailed up for the winter, as described for other forms. We continue to extend the stem from year to year, and to furnish its sides with fruit-branches only, keeping the degree of inclination before described. When they have attained two-thirds of their height, we bend them to an angle of 45 degrees. If they were inclined to this extent at once, it would stimulate too powerfully the growth of the lower shoots to the injury of the terminal one. When the stems have attained the height of the wall, the espalier is terminated, and the extremities must be treated in the manner directed for other peach trees that are completely formed.

No vacant space should be allowed upon the wall from one end of the stems to the other. This kind of espalier is commenced and terminated, as shown fig. 150, the same as the pear trees (pp. 110-12), and the other directions for pears in simple oblique cordon also apply to peaches. The advantages of this method of growing are as great in one case as in the other.

Trellis for Peaches in Single Oblique Cordon.—If a trellis be preferred, the following will be found the simplest and least costly form (fig. 153). It is made as follows:—

Fix at A a galvanised iron wire; pass it through the eyes of the iron hold-fasts, driven into the wall at B and C; then upon the round nails D and E; then bring it down, and through the hold-fasts F and G, then upon the nails H and I, then up and through the hold-fasts J and K, and fix it at L. Commence a new length of wire, fastening it at N, and continue as before to the end of the wall. These first lines of thick iron wire, fixed as we have directed, thirty inches apart, and inclined to an angle of forty-five degrees, will support the stems of the peach trees. The winter fastening, and the support of the fruit-branches, are provided for by two other lines of thinner iron wire placed on each side of the first, one about two and a half inches from the stem, the other about eight inches. One of these iron wires is fixed at a, then taken up to, and through, the hold-fast b, then round the nails c and d, then down and through the hold-fast e, round the nails f and g, through the hold-fast h, over the nails i and j, through the hold-fast k, and is fixed at l. A fresh wire is then fastened at the right of the thick wire, and continued to the end of the wall.

We use for fixing the wires the iron hold-fast and round nails described at page 106; and the tighteners, described at page 107, are placed at the points P, to tighten the wires. This kind of trellis costs (in

THE PEACH. 173

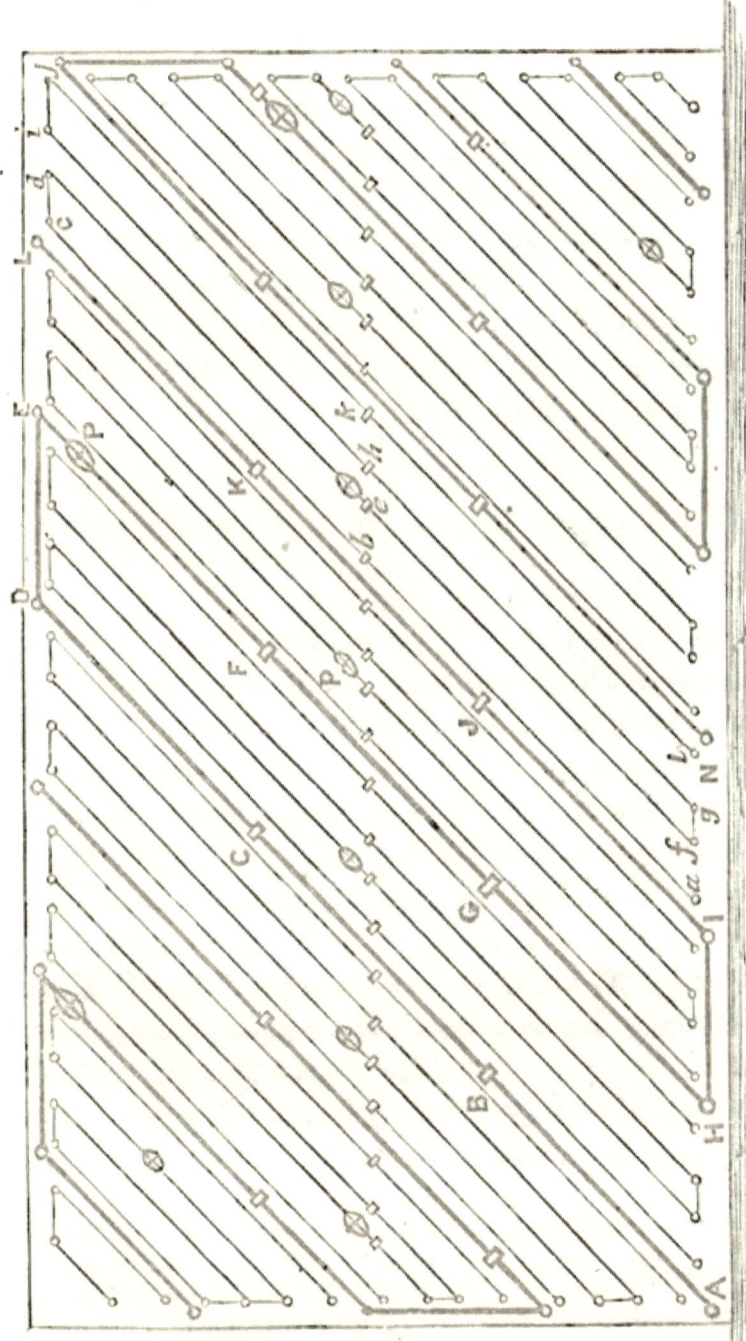

Fig. 153.—Iron Trellis for Oblique Cordon.

France) 10d. per square yard, not including the cost of fixing.

Pruning the Peach in Vertical Cordon.—When the wall is as high as twelve feet or more, it may be preferable to train peaches in vertical cordon, as described for pear trees (p. 119). In this case, the trees are planted two feet apart, and trained in the same way as pear trees of this form. If a trellis be used, it should be the same as the one just described, only the lines of wire must be vertical.

New Mode of Forming Fruit-Branches on the Peach.

The mode of training that we have already described is that which has been universally recommended and followed up to the present time.

For some years, however, we have been occupied with a new method, of which we did not intend to say anything until time had sufficiently proved its advantages. The new form* was first practised in 1847, by M. Picot-Amet, of Aincourt, near Magny, and a little later by M. Grin, sen., of Bourgneuf, Chartres, but with great improvements. We saw, in October, 1856, at M. Grin's, such striking results from this method, which had been practised by him for five years upon the same trees, that we do not hesitate to recommend

---

* This method is not altogether so new as may be supposed, for the leading principles of it are described by an English writer, and in the *Jardinier Solitaire*, published in 1712. But to M. Grin belongs the merit of re-discovery and directing public attention to it.

it in preference to all others. The mode of practising it is as follows:—

When the shoots of the successive branch extensions (fig. 154) have attained a length of about three inches, suppress only the buds behind, then the double or treble buds, so as to have only one bud at each point. The front buds must be preserved. At the

Fig. 154.—First Pinching of Shoots.

same time, these buds must be vigorously pinched off by the nail at A (fig. 154), above two of the lower leaves that are well developed. We do not count the small partially developed leaves, B, which often form a rosette at the lower part of the shoot.

Very soon after, we see new shoots springing from the foot of the leaves A A (fig. 155). These must be pinched as soon as they are two inches long, above their first leaf.

The irregular shoots will still spring from the base of the latter, as shown at A (fig. 156); but the season

Fig. 155.—Second Pinching of Irregular Shoot.

is too far advanced, and the sap acts with less force, so that the development is feeble, often attaining to a length of little more than an inch or two. Those at the top are the only ones that lengthen much. All of them are pinched above their first leaf, when they are about two inches long. If new shoots appear after this third pinching, they must be suppressed entirely. After the fall of the leaves, and at the winter pruning, these different shoots give place to a mass of branches, shown by figs. 157 and 158.

Fig. 156.—Irregular Shoots, pinched above the First Leaf.

The different pinchings that we have described have had the effect of gradually weakening the shoots, by concentrating the action of the sap towards the extension shoot of the principal branch. Thus all the shoots have given place to branches less vigorous, and which are covered with flower-buds.

In pruning these branches, we cut at the points A (figs. 157 and 158), so as to preserve only the fruit-buds of the lower part. During the following summer, the new shoots, which give rise to certain wood-buds

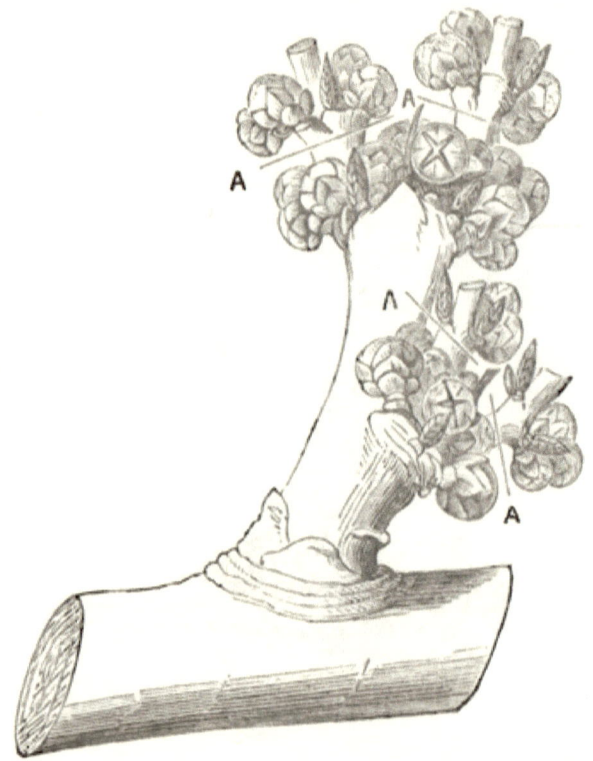

Fig. 157.—New Treatment of Fruit-Branch.

among the flower-buds, and which develop at the same time as the fruit, must be removed by pinching, as during the previous summer, and at the winter pruning must be cut still shorter, to concentrate the action of the sap towards the base, and to give rise to new fruit productions. The same mode of operating must be repeated from year to year.

The irregular shoots C (fig. 159), which spring very

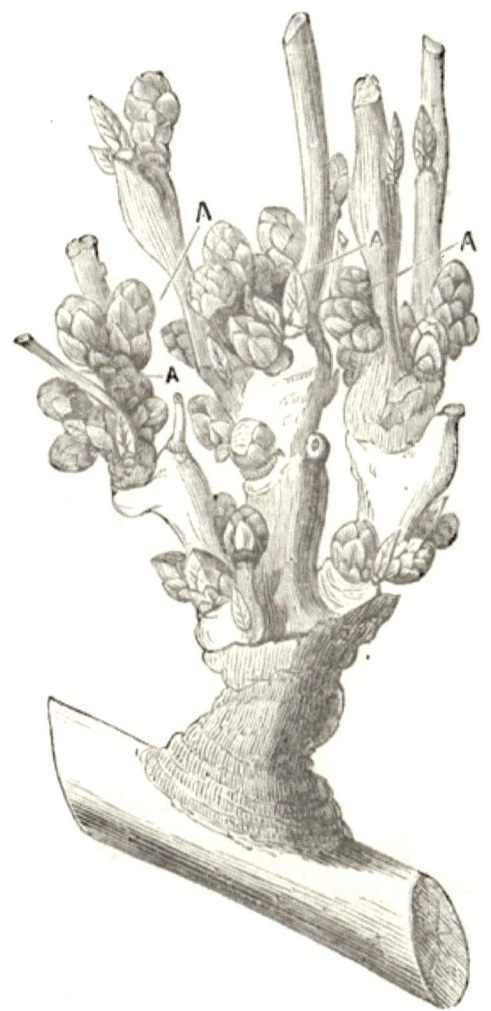

Fig. 158.—Another Fruit-Branch, after the same Treatment as Fig. 157.

numerously upon the shoots of the branch extension must be pinched off, those at the back entirely, and the rest pinched as soon as the second pair of leaves appear, and the eyes of the lower leaves are formed, the leaves at the base being alone retained. If this is performed too late, the pair of leaves at the base are drawn out by the lengthening of the shoot, and there will be at the winter pruning a branch resembling the figure 160. If the operation be performed too soon,

Fig. 159.—Irregular Shoot, pinched Short.

before the eyes of the lower leaves are properly formed, the shoot will wither, as shown at fig. 161. When, however, the operation is performed at the proper time, the shoot ceases to grow longer, and the lower pair of leaves remain at the base. At the time of winter pruning the branch is constituted, as shown at figs. 162 and 163.

It not unfrequently occurs that the irregular shoots develop so vigorously that, notwithstanding pinching, their stems continue lengthening and draw out with them the two lower leaves. To avoid this, M. Grin recommends an operation, of which we have proved the

Fig. 160.—Irregular Branch with One Shoot, pinched too late.

efficacy. As soon as one of the vigorous shoots appears, an incision about half an inch long is made with the point of the grafting-knife on one of its sides of attachment to the principal shoot A (fig. 159). The incision checks the growth of the shoot it produces, and the eyes of the lower leaves develop. After six or seven days they must be pinched as before directed. The result of these operations is shown at fig. 163. All the irregular shoots having been pinched the first time, there spring up on many of them one or two generations of shoots. These must be pinched above the leaf nearest to the base, the same as directed for buds proper. These operations result in branches formed as figures 160 and 162. We then cut them at B.

Fig. 161.—Irregular Shoot, pinched too soon.

Possibly these repeated pinchings practised upon the irregular shoots may result in a small branch covered only with flower-buds (fig. 164). If left to fructify, it will wither after the fruit is gathered, leaving a bare place. To avoid this, it will be advisable to suppress all the flower-buds at the winter pruning (fig. 165), then make a deep incision at A, penetrating below the insertion of the branch. We shall then see at the following spring, near the base, new shoots, better constituted, and which must be pinched short.

The same kind of incision, practised at the same period, at the base of the long irregular branch (fig. 160), will result in new shoots, but only in the

THE PEACH.   183

following spring. To facilitate this operation, prune the top of these branches very close. We take advantage of these new shoots to obtain a better formed branch.

The advantages resulting from this method of treating peach trees are as follow :—

1. We dispense with the summer fastening up of the shoots, and of the fruit-branches in winter; this allows of an ordinary trellis being used, such as for other fruit trees, of much less expense. Thus, for palmettes or other large forms we can use the trellis, fig. 94 (p. 106); for trees in an oblique cordon that described at page 116.

2. The winter and summer pruning is much simplified, and brought within the capacity of every gardener.

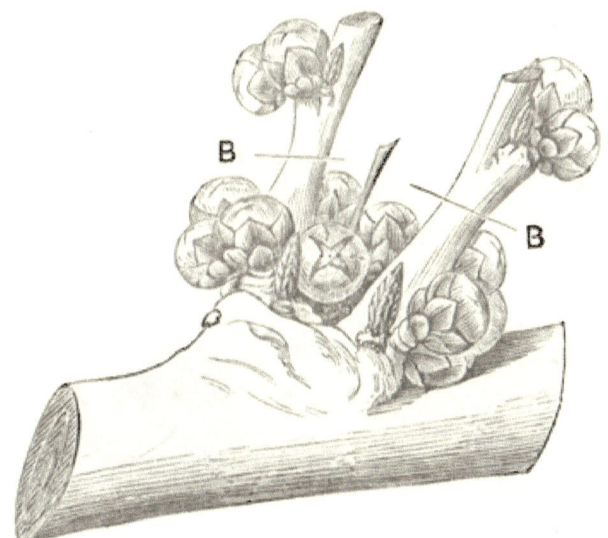

Fig. 162.—Irregular Branch with Shoot, pinched at proper time.

3. The fruit-branches may, by this method, be retained in front of the wood-branches, the front being

protected from the heat of the sun by the leaves. By the old method this could not be done, the fruit only growing on both sides of the branches.

4. The shoots and fruit-branches being kept much shorter, it is no longer necessary to leave between the wood-branches a space of 20 or 24 inches for the fastening up of the shoots and branches, an interval of twelve

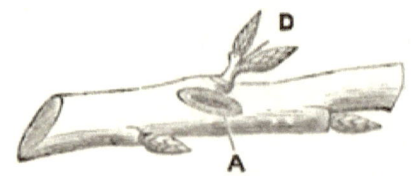

Fig. 163.—Small Irregular Branch with one Shoot.

inches being sufficient by the new method. By this means the number of principal branches may be doubled on a given surface of wall, and the quantity of fruit will be also double.

Fig. 164.—Irregular Branch, bearing only Fruit-Buds.

Peach trees trained as Verrier palmettes submitted to this modification will present the appearance of the pear tree figured page 96, and those of the oblique cordon, page 108.

True it is that the last advantage must be accompanied by an inconvenience in certain circumstances, when, for example, the trees are in the palmette or other large form. In this case it will be necessary to double the number of the principal branches. Now, as in general we are only able to obtain one set of these

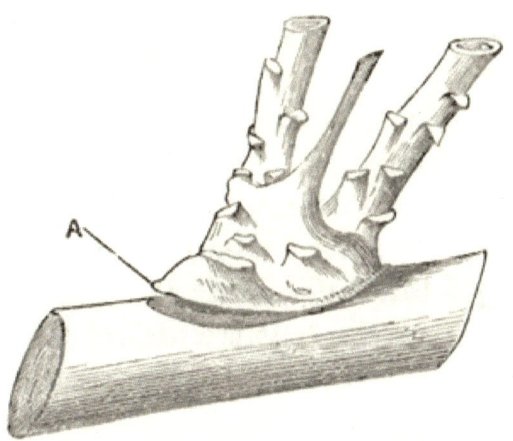

Fig. 165.—Being Fig. 164 deprived of the Buds.

side branches in a year, it follows that the wood cannot be made to cover completely the space it would occupy by the former method until after a lapse of sixteen or eighteen years, or half as long again.

The medium duration of the peach is twenty years; this new mode of management, therefore, loses its most important advantage in the case of large single peach trees. The new mode is of the greatest value when applied to peaches trained in vertical or oblique cordon. The trees need only to be planted at from thirteen to sixteen inches apart, instead of from 24 to 30 inches, as by the former method.

The plan is not only applicable to future planting, but quite as much so to trees already planted, be their age less or more. The method of proceeding is as follows:—

1. For trees that have been planted one year in oblique cordon, 30 inches apart, leave them to complete their second summer's growth, and in November shift them, bringing them within sixteen inches of each other.

2. Those that have been planted a year longer treat in the same manner.

3. Those trees that are of greater age, whatever be their form, cut the fruit-branches above the flower-buds nearest to the base; then, during the following summer, pinch back the shoots at the base of these branches. At the winter pruning suppress altogether the primitive fruit-branches; and the new branches resulting from the short pinchings must be cut off in the new manner.

The new method offers but a part of its advantages in the case of older trees, on account of the too great intervals between the principal branches. This may be remedied in the case of trees planted in cordon, by allowing a *gourmand* shoot to grow from the base of each peach tree, and obtaining from them a second series of stems, to be bent in the same manner as the rest. In the case of the larger trees the inconvenience will remain, but we shall obtain by the new method the advantage of doing away with the nailing and fastening up both in summer and winter.

We conclude by two observations which are of great importance for the success of this mode of training.

1. It will not be advisable to apply the shoot pinching until the trees have been planted one year. During the first year we must be content with the old method of pinching. At the winter pruning all the branches must be cut back above the bud nearest the base, and the shoot that springs from it must be pinched short. By this method the reforming of the tree is facilitated, and it furnishes the trees during the first summer with a great number of shoots.

2. The short pinching of the regular shoots ought to be commenced as soon as possible, that is, when the shoots have attained a suitable length. This must be followed up without interruption as shoots continue to lengthen. If this be delayed or be repeated too seldom, there will be too many shoots pinched at one time, which would stop vegetation in all parts of the tree, gum disease will set in, the fruit fall off, and, it may be, the tree will perish. These accidents, which make some persons condemn short pinching altogether, may be avoided by pinching soon enough and successively, in the course of fifteen or twenty days, so that the vegetation pursues its proper course unchecked from the first to the last pinchings.

Such is the mode of performing the new method of treating the fruit-branches of peach trees. By following it intelligently we obtain at once the valuable results we have pointed out, and which induce us to persevere in the new method.

THE PLUM.

Soil.—The soil most suitable for the plum is a rather open calcareous clay. As the roots do not strike downwards to any extent, a great depth of earth is not required. Sandy or damp soils are wholly unsuitable.

Choice of Trees.—It is usual to purchase the plants already grafted, but if it is preferred to graft afterwards, the method is as follows :—

Grafting.—The plum is grafted on stocks of the same species, generally obtained from shoots or suckers, which spring up in great numbers round the lower part of trees whose roots have been wounded from whatever cause. These shoots are planted out and then grafted. This mode of cultivation is, however, objectionable. We obtain in this way nothing but trees without proper roots, which do not take good hold of the soil, and exhaust themselves by the suckers that they throw out in great abundance; besides, they will not stand any degree of drought, and never acquire the dimensions of large trees. It is much better to use stocks obtained from plum stones, choosing from the most vigorous varieties.

The grafts are the same as recommended for the pear, budding being generally preferred as the most certain. Budding is generally performed in the month of July.

Varieties.—At the present time there are more than eighty varieties of the plum in cultivation, which may be divided into two groups: those eaten in their fresh state, and those dried as prunes. We give a list of some of the best varieties ripening at various periods.

NAMES AND SYNONYMES.	ENGLISH NAMES.	WHEN RIPE.	FORM.	ASPECT.
De Monfort.................	....	End of July. Au.	Stand. Espal.	E W S
De Monsieur	....	Begin. of Aug...	Stand. Espal.	W S
Gros hâtif.				
Reine-Claude ordinaire ...	Green Gage...	End of August .	Stand. Espal.	E W S
Verte-et-bonne.				
Reine-Claude abricot, vert.				
Reine Victoria	Victoria	End of August .	Stand. Espal.	W S
Petite Mirabelle	....	Begin. of Sept..	Stand. Espal.	
Reine-Claude r. Van Mons	....	Middle of Sept..	Stand. Espal.	W S
Reine-Claude violette......	Purple Gage..	Middle of Sept..	Stand. Espal.	W S
Reine-Claude de Bayay ..	Reine-Claude de Bavay	End of Sept......	Stand. Espal.	E W S
Coe's Golden Drop	Coe's Golden Drop	Begin. of Oct....	Stand. Espal.	E
Waterloo.				
De la Saint-Martin	....	End of October	Standard	
*D'Agen	....	Begin. of Sept..	Standard	
Robe de Sergent.				
*Washington................	Washington..	Middle of Sept..	Standard	
*Pond's Seedling	Pond's Seedling	Middle of Sept..	Standard	
*Conestche d'Italie	....	End of Sept.....	Standard	
Fellemberg.				
Prune suisse.				
*Sainte-Catherine	Catharine......	End of Sept.....	Standard	

* Those marked with an Asterisk (*) are Plums for Drying.

A Selected List of Plums for General Cultivation in England.—The sorts are arranged in the order of their ripening, which is desirable and necessary in writing a list of fruits. We are indebted to the French for the greater number of the sorts of plums now in cultivation in this country. The plum does not thrive so well trained as an espalier. In the list those marked with an asterisk do best on a wall, the rest are standards.

Names of Plums for Dessert.	Names of Plums for Cooking.	Plums for Walls.
*July Green Gage.	Early Orleans.	July Green Gage.
Peach.	Mrs. Gisborn's.	Coe's Golden Drop.
*Perdrigon Violet.	Goliath.	Green Gage.
Hâtive.	Prince of Wales.	Purple Gage.
*Green Gage.	Victoria.	Italian Quetsche.
*Purple Gage.	*Diamond.	Blue Imperative.
*Jefferson.		Ickworth Imperative.
Kirke's.		
	For Preserving.	Plums for Standards.
*Coe's Golden Drop.	Green Gage.	Early Prolific.
*Reine-Claude de Bavay.	*White Magnum Bonum.	Early Orleans.
		Orleans.
Coe's Late Red.	Washington.	Prince of Wales.
Cooper's Large.	Wine Sour.	Victoria.
Late Orleans.	Damson.	Pond's Seedling. (Very good.)
	*Diamond.	Coe's Late Red.
	Autumn Compôte.	Damson.

TRAINING AND PRUNING.

The plum is cultivated either in the garden or orchard. It is usual to plant in gardens as standards, not as espaliers, but this is a mistake; espaliers produce a better quality of fruit than standards. The best forms for standards are the pyramid and contra double espalier in vertical cordon; and for espaliers, the Verrier palmette, simple oblique cordon, and vertical cordon. For trees planted in the orchard the tall standard form is the best. We proceed to inquire what treatment these four forms of the plum require.

THE PLUM.

Training the Plum in Pyramid Form.

Formation of the Wood.—The mode of proceeding in this case is the same as for the pear. Trees one

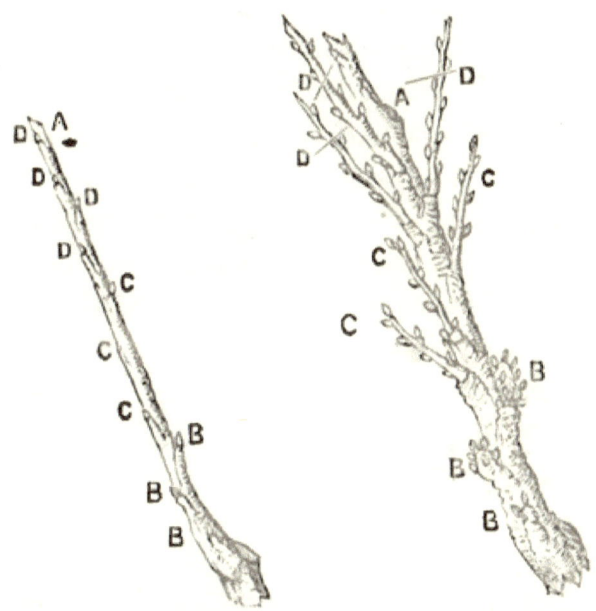

Fig. 166.—First Year's Pruning for forming Fruit-Bud Branches.

Fig. 167.—Second Year's Pruning of Fruit-Branches.

year old from the graft are placed at the same distance apart. We have nothing to add to the directions previously given for the pear.

Management of Fruit-Branches.

First Year.—Take one of the lateral branches of the pyramid (fig. 166). This branch presents at the spring following its first year nothing but wood-buds. During the following summer the branch which has been cut at A, in order to make it develop all its buds, including those at the base, has transformed each of them into shoots more or less vigorous as they are nearer the extremity of the branch or further from it. Those at the lower part B only develop very minute shoots; those at C, growing about the middle, attain a length of from two to four and a half inches; lastly, those at D will perhaps grow to from eight to twenty inches in length. These last, with the exception of the terminal shoot, are pinched off when they have attained a length of four inches, in order to transform them into fruit-branches and to favour the extension of the terminal shoot. We suppress besides all the double and treble shoots, preserving the most feeble or most vigorous, according as it is desired to obtain fruit-branches or an extension of the branch.

Second Year.—At the spring which follows, this branch presents the aspect of figure 167. The very small branches at the base B support a group of flower-buds, in the centre of which is a wood-bud which will form the extension of this small fruit-branch. These small branches must remain untouched. Others that are larger, C, and which bear flower-buds about the middle part, and wood-buds at the top, must be cut back. Those at D that are too long and too vigorous must be shortened by complete or partial fracture, as they are more or less vigorous, in the manner we have explained for the pear trees. These cuttings are necessary in order to obtain new branches towards the base, for it must not be forgotten that the fruit-branches of stone fruits only bear fruit once. During the following summer we pinch again those of the new extensions that have attained a length of two or three inches.

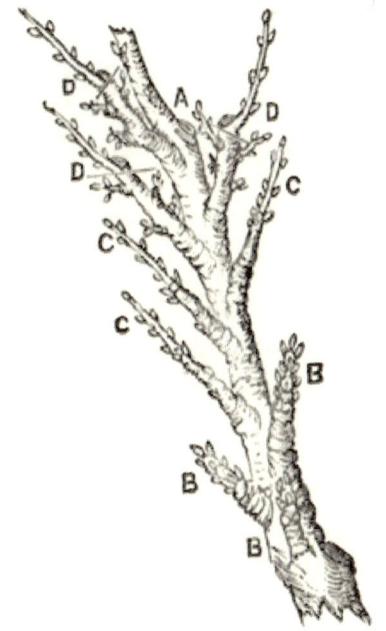

Fig. 168.—Third Year's Pruning of Fruit-Branches.

Third Year.—At the third spring the branch is constituted as figure 168. It will be observed that the small branches B and C have grown a little, and those at D have shot out new branches; most of these last must be cut back a little in order to diminish the number of the blossoms, and to prevent the too great lengthening of the shoots. The same operations must be continued from year to year, carefully cutting back, not only the branches D, which will continue to lengthen, but still more C and B, in order to develop new replacing branches towards the base.

K

Plums as Double Contra Espaliers in Vertical Cordon.

The same methods are used as directed for pears of the same form.

Plums as Espaliers—Verrier's Palmette.

In this case also the same process is adopted as for pears. The fruit-branches must be treated as described for plums in the pyramid form. Space between each tree should be allowed, so as to afford about twenty square yards upon the wall for every tree.

Plums in Single Oblique and Vertical Cordon.

These are particularly suitable forms for the plum. The same operations are required as for the pear. The vertical cordon is suitable under the same circumstances as for the pear, viz., for walls at least five yards high.

Plums as Tall Standards.

This form is suitable only for orchards. The same mode of treatment as the pear.

THE CHERRY.

Soil.—The cherry prefers a dry to a damp situation, and a sandy, and, most of all, chalky soil of medium consistency.

Grafting.—The cherry is grafted on two kinds of stocks, the St. Lucie plum and wild cherry. The St. Lucie plum is preferable for low-stemmed trees of whatever form. These trees both look better, and accommodate themselves more easily to all situations. The stocks from the wild cherry produce the most vigorous trees, but are more liable to gum, and require a better soil. They are only suitable for tall standards.

The shield graft is the kind generally used, except for the wild cherry stock, which may be grafted with the cleft or crown graft, if too old for the shield graft. The St. Lucie is shield grafted at the beginning of September, the wild cherry in August.

Varieties.—About eighty varieties of the cherry are cultivated; we recommend the following selection for coming to maturity at successive periods :—

[The cherry thrives best in a light sandy loam upon a dry sub-soil, and the stock most suitable for English culture (except for pots in orchard houses) is that raised from seed from the wild black cherry. The cherry is grafted in March, or more commonly *budded* in July or August. For standards they are best budded near the ground, as they then grow a tall straight stem.—ED.]

NAMES AND SYNONYMES.	WHEN RIPE.	FORM.	ASPECT.
Bigarreau de mai	End of May	Espalier	S
Angleterre hâtive................	Beginning of June	Espalier	E W S
May Duke.			
Belle de Choisy	June	Espalier	W S
Doucette.			
Griotte de Chaux	End of June	Espalier	W S
Griotte d'Allemagne.			
Royale cherry-Duke............	End of June	Stand. Espa.	W S
Dowton...............................	Beginning of July	Stand. Espa.	W S
Noire de Prusse..................	Beginning of July	Stand. Espa.	W S
Reine Hortense	Beginning of July	Stand. Espa.	E W S
Monstrueuse de Bavay.			
Belle de Sceaux..................	End of July	Stand. Espa.	E W S
Belle de Chatenay.			
Marello de Charmeux	End of Aug. to end Oct.	Stand. Espa.	E W E

[The following is a selected list of cherries suitable for English cultivation. They will all do well in the open ground as standards or espaliers, but the dessert kinds are improved by being grown against a wall; they will be earlier, and the flavour better.—ED.]

DESSERT.

June or middle of July.
    Belle d'Orléans.
    Early Prolific.
    Monstrous Heart.
    Mammoth.
July to middle of August.
    Knight's Early Black.
    May Duke.
    Waterloo.

Black Eagle.
Elton.
Jeffry's Duke.
August.
    Florence.
    Coe's Late Carnation.
    Late Duke.
    Kentish.
    Morello.

THE CHERRY.

ORCHARDS.

May Duke.
Elton.
Black Hawk.
Hogg's Black Gem.

Mammoth.
Late Duke.
Knight's Early Black.
Early Prolific.

PRUNING AND MANAGEMENT.

The cherry resembles the plum in its mode of growth and fructification; we have, therefore, little to add to the directions given for the plum.

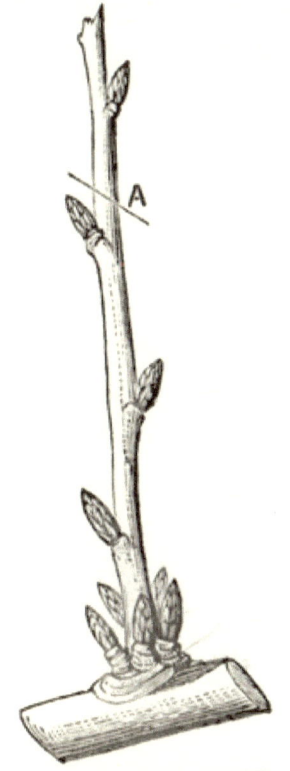

Fig. 169.—Fruit-Branch of Cherry a Year Old.

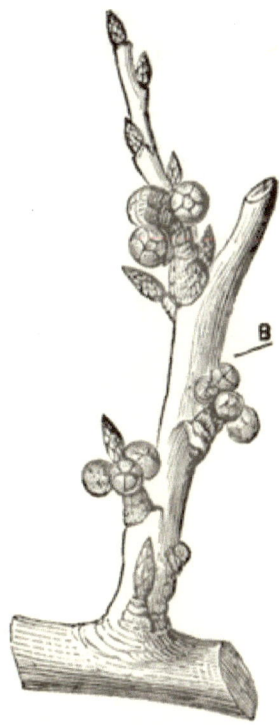

Fig. 170.—Fruit-Branch Two Years.

The cherry is grown chiefly in three forms: the

standard for orchards; the pyramid, the *double vertical cordon*, and espalier for the fruit-garden; *Verrier palmette* and *oblique simple cordon* are the forms we recommend for espaliers, to which may be added the *vertical cordon*.

The management of the wood and the fruit branches is the same as for the plum. Thus the shoots which grow upon the last extension of the wood-branches, being pinched at three or four inches in the summer, produce, as the result, a branch resembling fig. 169. At the winter pruning these are cut at A. After another year's growth these branches present the aspect of fig. 170. They are next cut at B, in order to force back the sap to the lower part, to obtain new fruit-buds, and so on from year to year.

THE APRICOT.

Climate and Soil.—The apricot ripens in all climates of France; but as it blossoms early, its fructification is often destroyed by the late frosts of spring. Thus its cultivation as a standard is not profitable further north than Paris. In a colder climate than this it must be trained exclusively as an espalier [wall tree]. This is an unfortunate necessity; for, contrary in this respect to its allied species, the flavour of espalier apricots is much inferior to that of standards. The soil best suited for the apricot is the same as for the plum.

Grafting.—The apricot is always multiplied by shield grafting (budding) upon plum stocks.

Varieties.—The apricot has about twenty varieties, of which the following are the best:—

NAMES AND SYNONYMES.	WHEN RIPE.	FORM.	ASPECT.
Musch	Middle of July	Stand. Espa.	E W S
Montgamet	End of July	Stand. Espa.	E W S
Gros common	Beginning of August	Stand. Espa.	E W S
Royal	Middle of August	Stand. Espa.	E W S
Pêche	End of August	Stand. Espa.	E W S
De Nancy.			
Beaugé	Beginning of Sept.	Stand. Espa.	E W S

A selection of the most desirable apricots for cultivation in England :—

> Moorpark Apricot, ripe beginning of September. The best bearer and most desirable for general cultivation.
> Roman Apricot, an excellent bearer when the tree gets a little aged, or arrives at maturity ; ripens about the middle of August.
> Orange Apricot, ripens in August.
> Hemskirke Apricot, ripe about the second week in August.

The above apricots succeed well on a wall on any intermediate aspect from east to west.

[The apricot is one of our principal English wall fruits, as it comes into bearing earlier, and ripens its fruit more quickly than any other tree that requires the assistance of a wall. It *will not* ripen its fruit even in the south of England (to depend upon for a crop) either as an espalier or a standard. It requires the best situation against a wall, especially in the midland and more northern parts of England. The different kinds of apricots are propagated by *budding*, in July or August, on plum stocks, two of the best of which are the Muscle plum and the Brussels plum; the first is more generally used.—ED.]

PRUNING AND MANAGEMENT.

The apricot, as before stated, is best cultivated as a standard, where the climate or situation is such as to insure the preservation of its fruit from the frosts of early spring. In the fruit-garden it may be trained in the goblet form, or better as an espalier. The form may be *Verrier palmette*, or *simple oblique cordon*. Choose the best and most sheltered situation in the

THE APRICOT. 201

garden, and cultivate against a trellis (fig. 171). The trellis C is fastened to the supports B, placed six feet apart, and being at least seven feet in height; any of the forms of espaliers may be adopted that have been previously described. About the middle of February fasten straw matting, A, behind the trellis, extending from the top to the bottom, make this quite secure, and then fix a roof of straw or matting, E E, at the top, projecting about two feet. This shelter will protect

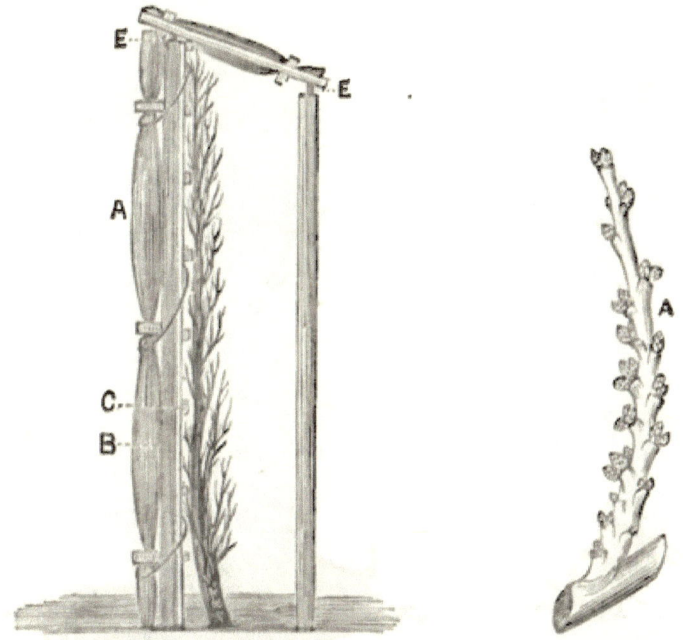

Fig. 171.—Cover for Apricots in Contra Espalier.

Fig. 172.—Apricot Fruit-Branch before Pruning.

the trees from cold. They may be taken off at the end of May, and the apricots will grow and ripen as well as upon standards.

The management of the fruit-branches is the same as

K 3

that of the two preceding species; thus the branch A (fig. 172) is the result of pinching practised upon the shoots that spring from the new extension of one of

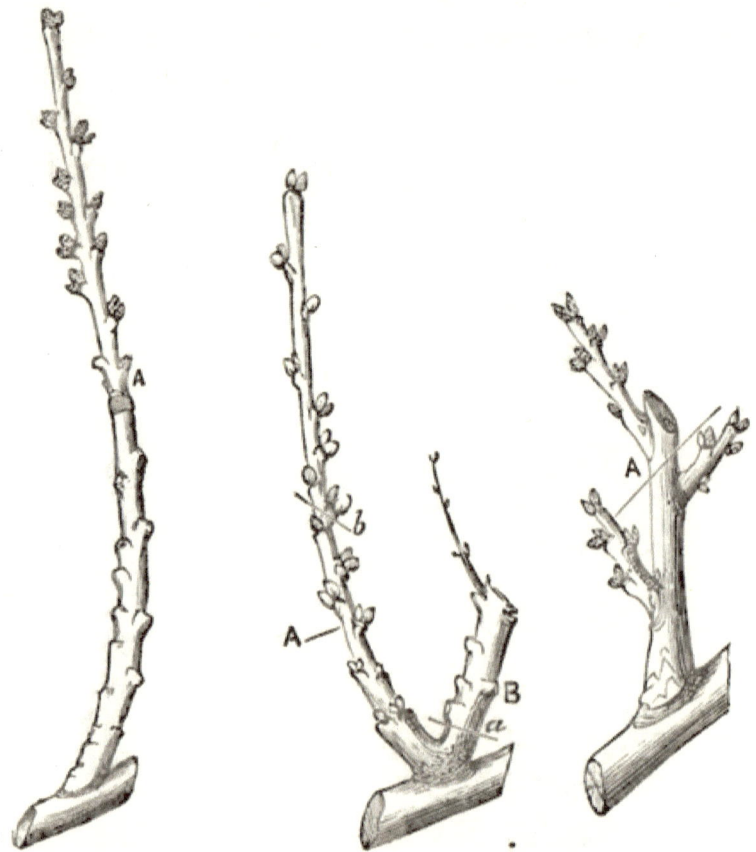

Fig. 173.—Apricot Fruit-Branch, Pruning neglected.

Fig. 174.—Fruit-Branch, Year after First Pruning.

Fig. 175.—Another Fruit-Branch, Year after First Pruning.

the branches. If left entire until the winter pruning, they will fructify and send out a new terminal branch (A, fig. 173); if this is left entire, it will lengthen itself again, and thus year after year become weaker until it will finally disappear, leaving a vacant space.

It will be necessary to cut back the branch at A (fig. 172), preserving a sufficient number of flower-buds to force the sap towards the base, in order to obtain new fruit-branches. The result of this operation will be a new branch, like the figures 174 and 175. In the first case (fig. 174) the primitive branch B at *a*, and the new fruit-branch A at *b*, are cut in order to obtain again the same result. On fig. 175 we suppress one of the small branches at A to make a new fruit-branch spring from the base, and so on every year.

RESTORATION

OF

BADLY-TRAINED AND AGED TREES.

Few trees are trained in the careful manner prescribed in the foregoing pages. It is not, therefore, a matter of surprise that a great number fail to be so productive and profitable as probably would be the case if they were properly trained. Are they then to be destroyed and replaced by new ones? No. In most cases they may be restored by suitable operations, if not to perfect symmetry, yet to as much regularity and fertility as they are capable of.

All fruit trees, however carefully managed, become weak and comparatively barren when they have exceeded a certain age. Trees in this condition ought not in all cases to be destroyed, for they may for the most part be renovated. As profitable results may be obtained by this means sooner than by planting a new orchard, the subject is one of much importance.

Restoration of Badly-Trained Trees.

We shall treat standards and espaliers separately.

Standards.—The lower side branches of standard trees intended for the pyramidal form are generally pruned too short, and those immediately above them too long. The result is, that nearly all the sap flows towards the top of the tree, leaving little for the lower part, and the branches towards the base are arrested in their growth before they have attained their proper

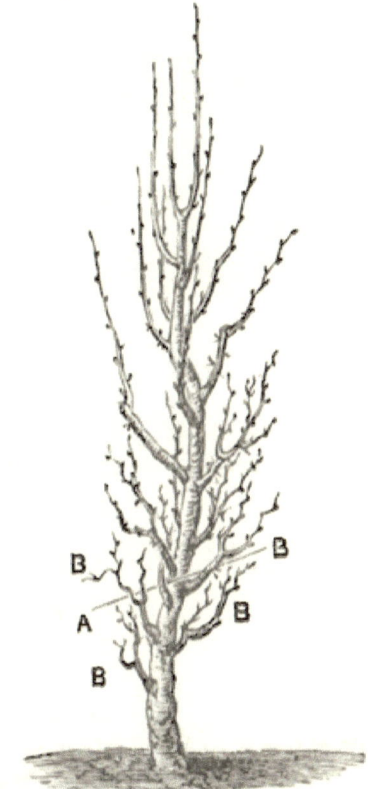

Fig. 176.—Restoration of a Young Pyramidal Tree.

length; owing to this cause, the lower branches bear a too great abundance of fruit, which weakens them, and gradually they die off, the tree running to head.

If the trees are not higher than five or six feet (fig. 176), and retain sufficient vigour, it is best to cut them at about 20 inches above the ground, and to cut

Fig. 177.—Restoration of an Old Pyramidal Tree.

back the lateral branches, B, to the stem, and afterwards proceed in the same way as for young trees intended for the standards of the pyramidal form, from their first start.

But when the tree has attained a height of twelve or fifteen feet, as fig. 177, and the base of the stem is still provided with a certain number of branches, two-thirds only of the total height should be cut away; and the branches situated immediately below cut back to about one and a half inches from the stem. The branches quite at the base must be left entire. The branches situated between these two points must be cut so that their highest points maintain an oblique line from the extremities of the lower branches to those of the branches at the top of the tree. It is also necessary to make incisions upon the stem, either to determine the formation of new lateral branches, where they are deficient, or to strengthen feeble branches or those that have been newly grafted upon the stem by approach on the Richard side graft (pp. 5 and 17), where incisions alone have not succeeded in developing new shoots. If, however, the lower branches are too feeble, relatively to those situated at the point where the stem is cut, and also if their number is insufficient, it may be necessary to cut them away altogether, in order to develop a new series of side branches.

During the following summer the growth of the lower branches must be accelerated by pinching those at the summit, with the exception of that, however, which has been retained for lengthening the stem. At the winter pruning the lower branches should be left almost entire, and the others cut back successively, giving a length of six inches to those at the top, and of twelve to the central one which continues the stem.

During the following summer the sap is again forced back towards the lower part by means of continued pinchings, and the tree will then shortly begin to assume the desired pyramidal form.

Espaliers.—It is necessary to make a distinction between the treatment of *pépin* and stone fruit.

In proceeding to restore pear or apple trees, of whatever age, provided they are sufficiently vigorous, to a regular form, we seek among the various ramifications at the base three branches, one to form the new stem, and the other two for the first principal side branches. All the other branches are removed, and the two side ones cut back to about twelve inches. The stem is cut immediately above the place where the second stage of side branches ought to spring. Afterwards the treatment is the same as for palmettes, previously described.

If the trees contain no branches suitable for the purpose, they must be cut at twelve inches above the ground to make them develop three shoots proper for commencing the wood of the palmette.

These methods of renovating *pépin* fruits will almost always prove successful, because they possess the property of developing new shoots upon even the most aged branches; but this does not apply to stone fruits, and particularly not to peaches. Their restoration is therefore much more difficult, however vigorous they may be.

Restoration in this case can only be assured when

there are young branches so situated that they can be used for forming the new wood, or in their absence a bud or a small shoot at the base. In the former case we only reserve the branches useful for forming the new wood; in the second case we cut off the stem entirely above the leading or small shoot, and when this is developed make it the base of a new tree.

Renovation of Aged Trees.

Whatever care be taken in the pruning of fruit trees, it will not fail to happen in the course of time that there will be found knots close to the fruit-branches, caused by cutting, and the successive removals of those branches. These protuberances are serious obstacles to the free circulation of the sap from the roots to the bud, and from the shoots to the roots; a state of disease quickly sets in, and the tree soon perishes.

If the tree is taken in hand before it falls into a state of complete decrepitude, it is almost always possible to rejuvenate it, and restore it to its original vigour, especially if it is an apple or pear. But success is far less certain in stone fruits, most difficult of all in the case of peaches, which scarcely ever send out new shoots from the old wood.

Trees of Pyramid Form.—The essential point in renovating trees is to concentrate all the little sap that the tree has to dispose of upon a smaller extent of stem and branches, to make it put forth new shoots, and by the same means a new set of roots. It will be sufficient for such a tree to cut the stem half-way up, and also to cut the lateral branches, increasing their length to-

Fig. 178.—Pyramidal Pear Tree Renovated.

wards the base so as to preserve the pyramid form.

Those at the base should be cut at two feet from their spring, and the upper one at six inches. If the tree is an apple or a pear of large size with thick hard bark, it will be better, instead of reckoning upon the growth of new terminal shoots for the stem and branches, to crown graft each of the extremities, which will develop more vigorously than mere shoots from old trees.

At the end of the year the rejuvenated pyramid will resemble fig. 178. During the first years which follow, in order to favour the growth of the lower branches, it will be necessary to cut short those at the top, and to pinch, during summer, their terminal shoots.

Espaliers.—Espaliers in the palmette form are cut at one-half the length of their stem, and the side branches cut back, those at the top to six inches, those at the base one-half their length, and the intermediate ones so as not to pass their proper line of growth. A crown graft is also placed at the top of the stem and at the extremities of each side of the branches. Lastly, the lower branches are favoured in their growth by cutting short the upper ones and pinching their shoots.

To further insure success it will be well to take off the old outside bark of the entire tree with a plane, and then cover the whole with a coat of lime-wash, applied hot. This will stimulate the vital energies of the tree, and facilitate the growth of new shoots.

It will also be an advantage at the close of the third year to make a circular trench three feet from the foot of the tree, three feet wide, and thirty inches deep, and fill it with fresh earth enriched with manure. If some of the old roots are met with in digging they must be left untouched.

We ought to observe that the operations now described for the restoration and regeneration of trees have lost much of their importance since the introduction of the method of growing fruit trees in oblique and vertical cordon. These forms allow of the maximum product of standards and espaliers being obtained about the sixth year after plantation, while fourteen or sixteen are required for restored or rejuvenated trees of the larger forms; hence it follows that it will generally be more advantageous to substitute for these operations a new plantation in cordons. In the case of all the trees in a garden or orchard requiring these operations to restore and rejuvenate them, we proceed thus:— Remove altogether the old trees upon a third part of the ground where they are in the worst condition, and replace them by occupying the ground with cordons. When these begin to fructify we proceed in the same way upon the second third of the ground, and some time after on the rest of the space. All the garden or orchard will be thus restored without ever being deprived of fruit.

GENERAL DIRECTIONS

FOR

THE FRUIT GARDEN.

IN addition to the directions already given there are certain other indispensable points to be observed in order to assure a vigorous growth of the trees. These relate to the annual cultivation of the borders on which the trees are planted, the protection of the trees from the late frosts of spring, or the too great heat of the sun in summer.

The cultivation of the borders on which the trees grow is of much importance. The digging renders the soil permeable at all times to the action of the atmosphere, and frees it from weeds; a sufficient quantity of fertilising matter is supplied by the manure, and the trees are preserved from drought.

Digging.—This should not be too deep, to avoid injuring the roots, particularly of trees grafted on quince, plum, or paradise stocks, which always develop themselves more superficially than others. In the latter case, instead of using the spade, it will be better to employ the fork or hoe fork, fig. 179, which is less liable to cut the roots. The digging should be performed every year immediately after pruning.

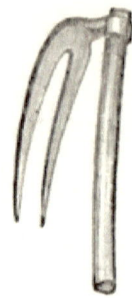

Fig. 179.—Hoe Fork.

It is a common practice to grow vegetables upon fruit-tree borders; but it is a bad one, for the numerous diggings to which the ground is subjected will constantly expose the roots of the trees to mutilation. Besides, the vegetables absorb nearly all the nutriment of the soil. At all events, we should confine the practice to the growth of such vegetables as least exhaust the soil, such as salads. Above all, avoid planting cabbage.

Manuring.—It is not desirable to manure fruit trees very abundantly until they have attained their intended dimensions, otherwise the production of fruit will be retarded. Some manure every three years, but the practice is wrong, for it compels the use of too much manure at a time, and the fruit contracts a bad flavour, and stone fruits, especially peaches, are liable to the gum disease from this cause. It is better to manure a little at a time every year.

In clayey soils it is usual to employ horse and sheep manure rather fresh, and in lighter soils that of the cow. We prefer, however, crushed bones, woollen rags, waste wool, hair, or feathers. This forms very powerful manure, decomposes slowly, and the effect is greatly prolonged. It is better for the trees than ordinary manure. It will be sufficient to repeat such manuring every seven or eight years. Any surplus manure that remains may be dug into the border occupied by the roots.

To Counteract Excessive Dryness of Soil.

The operations consist in watering, covering, and keeping the surface of the soil light and open.

Watering.—The great heats of summer render frequent waterings necessary, most of all in light soils, and for trees recently planted. In order to prevent the hardening of the surface of the soil, it is desirable to cover the foot of the tree with litter. Young trees ought to be well watered every eight days. Liquid manure is preferable to water when it can be had, as it stimulates the activity of the vegetation. Watering should always be performed after sunset.

Lightening the Soil to the depth of two inches immediately it begins to dry and harden. This will take the place of watering for trees after their first year of planting. This practice is most resorted to upon strong lands.

Covering.—This produces the same result as the last, and consists in covering the soil round the trees with dried leaves, decomposed straw, fern, &c., to the depth of two or three inches. May is the proper month for this. It is most employed for light soils.

PROTECTION FROM THE LATE FROSTS OF SPRING.

The late frosts, snow, hail, and cold wet weather of spring are extremely injurious to fruit trees, and notably so to stone fruits. We shall point out what may be done to counteract this, considering espaliers and standards separately.

Espaliers.—The coping of walls against which fruit trees are planted often projects ten or twelve inches; insufficient to protect from cold, it becomes positively injurious towards the end of May, by depriving the trees of the full action of the warm and genial summer rains. It is, therefore, better to let the copings project only about four inches, and to protect the trees thus :—

For walls without trellis we fasten a number of wooden rods or supports (fig. 180), projecting about

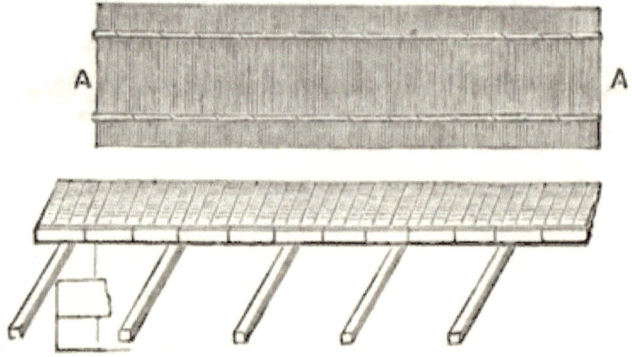

Fig. 180.—Protection for Espaliers.

two feet, and inclined to an angle of about 30 degrees. In February, when the trees begin to vegetate, we fasten to these supports the straw matting made into the form of hurdles (A, fig. 182), by means of two wooden rods before, and two behind, and the whole fastened together with iron wire. These protectors are kept up until the end of May, when the fruit begins to set.

Upon walls covered with trellis a wooden horse (fig. 181) is substituted for the wooden rods. These protectors are indispensable for stone fruits; but apples and pears will be much better for them, particularly

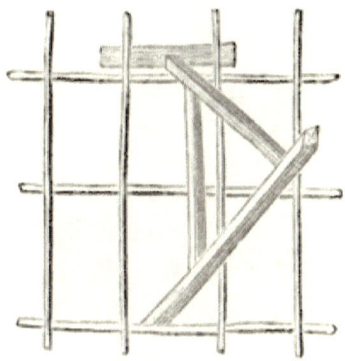

Fig. 181.—Matting Horse.

those which have a west or north aspect, in damp localities. In this case the protection should be put up in June, and kept in use till the end of September. They prevent injury from fogs and cold rains of summer. These weather protectors should not project more than about sixteen inches. They will be sufficient to protect the trees from injury from cold when the temperature is as low as one degree below freezing; but are insufficient against colds of two or three degrees, which too often destroy the fruit crops of our gardens. Let us see what can be done to protect the trees against these more severe frosts.

For this purpose we place a rod of wood upon the lower part of the top covering at B, fig. 182, supported

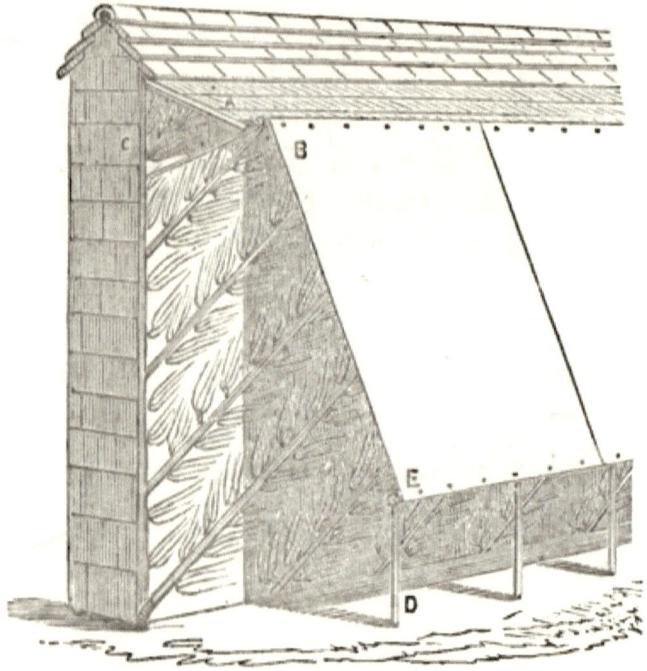

Fig. 182.—Protection for Stone Fruit Trees.

by the extremity of the house which forms the projection. We then drive a line of stakes, D, into the ground, about 30 inches high, placed 5 feet apart, and 5 feet in advance of the wall. We fix at the top of the stakes a cross piece, E; then stretch a cloth from B to E. The cloth may be of the coarse canvas used for walls before papering. Its durability will be increased by soaking it in linseed oil. The canvas allows of light passing through it sufficient for the trees, and will protect the vegetation from the most severe of the late frosts of spring. We secure by this means almost as many fruits as there have been blossoms. The space underneath is sufficient to allow the gardener to pass, and to perform his various operations of disbudding, pruning, and pinching without hindrance. This covering and the straw hurdles are laid aside for the year towards the end of May, when the fruit is nearly set, and there is nothing more to fear from late frosts.

Standards.—It is much more difficult to protect standard fruit trees. The only practicable method of doing so consists in fixing upon the branches, immediately after the pruning, handfuls of dry fern or straw, so that each branch may be protected to its full extent. Or the tree may be covered all over with canvas, like that just described. To protect cordons, double or vertical, we can stretch the canvas horizontally at the top, and fix it with wire, to be taken off towards the end of May.

Shading.—Espaliers, and especially stone fruits, are exposed in all their green surfaces to such a powerful evaporation, that the roots cannot supply the loss of humidity that takes place. Besides this, their position deprives them in a great measure of the benefits of the night dews, little as they may be, during the great heats of summer.

Unless protected in some way, many trees will then perish, as we may say from sunstroke. To prevent this, they must be watered freely, and the foliage be well syringed three times a week.

The heat of the sun is not less injurious to the bark of the stems of espaliers, most of all to the part not protected by the leaves. It hardens the bark so that it loses its elasticity, does not allow the tree to expand, and hinders the circulation of the sap, by compressing the sap vessels. The bark is sometimes destroyed in this way, and falls off, leaving the wood bare.

Fig. 183.—Wood Shade for Stems.

To prevent this the more exposed part of the stem may be covered with a shade made of wood like fig. 183. The upper part may be protected by a coating of half whiting, half clay, mixed with sufficient water to form a thick paste.*

---

* The injury here referred to applies mostly, if not entirely, to the climate of France. In England fruit trees are not often injured by too much sun.

GATHERING

AND

PRESERVATION OF FRUIT.

Gathering of most fruits should be performed before the fruit is quite ripe; the quality and flavour will be better for it. But it will not do to anticipate their maturity for more than eight days for pears and apples, and one day for peaches, apricots, and plums. Cherries should only be gathered perfectly ripe.

Pears and apples which are not ripened before winter must be gathered in October, or when the vegetation of the trees ceases. Whatever be the kind of fruit, it should only be gathered when quite dry, and on fine clear days. The fruit has then its finest flavour, and will keep much better.

The best method of detaching the fruit is to gather it carefully one by one with the hand; various contrivances have been devised, more or less ingenious, for gathering those at the top of the trees, but all of them are liable to injure the fruit, and it is better to reach them by a ladder.

As the fruits are detached, they should be placed in a large shallow basket, lined at the bottom with moss or dry leaves. Not more than three layers should be disposed in one basket, and each layer should be kept separate by leaves. The fruit must be taken immediately under cover.

Preservation of fruits applies mostly to those fruits which only ripen during winter. The object is :—

1. To preserve them from frosts, which completely disorganise them.

2. To so manage that the ripening takes place gradually, and is prolonged, for a portion of the fruit, until the end of May. The complete or partial success of this depends upon the construction of the fruit-room, or place where the fruit is kept.

The Fruit-house.—Experience proves that the fruit-house or fruit-room affords the most satisfactory results which fulfils the following seven conditions :—

1. An equal temperature at all seasons.

2. A temperature eight or ten degrees above freezing.

3. Complete exclusion of the light.

4. Absence of all communication between the fruit-room and the exterior atmosphere.

5. The place should be dry rather than damp.

6. Such an arrangement as prevents, as much as possible, the fruit being injured by the pressure of its own weight.

7. A northern aspect, on a very dry soil, slightly elevated.

These are the arrangements of a fruit-house that we think fulfil all the required conditions. The size, of course, must depend upon the quantity of fruit. That of which we now give a plan (figs. 184 and 185) is about sixteen and a half feet in length, thirteen feet wide, and ten feet high. Eight thousand fruit may be stored in it, allowing four square inches to each fruit.

The floor is twenty-eight inches below the surrounding ground; if it is very dry it may be three feet lower. This arrangement allows of the atmosphere of the fruit-house being more easily protected from the exterior atmosphere. To prevent the wet from drain-

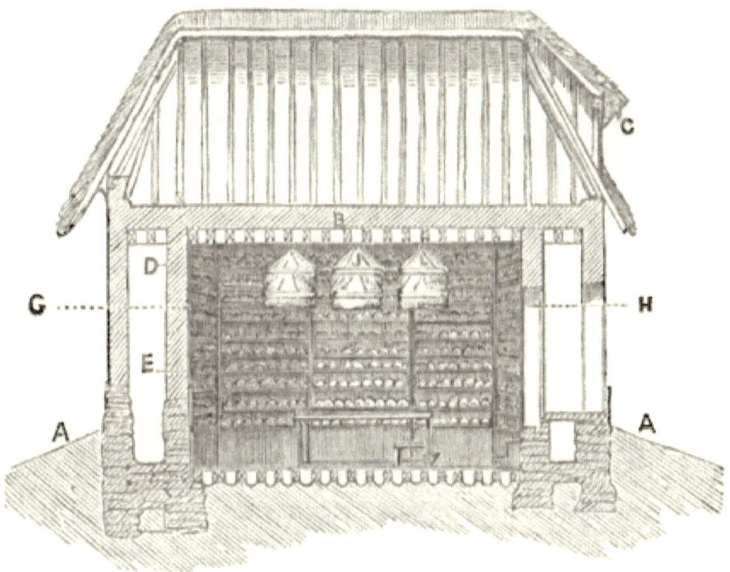

Fig. 184.—Elevation of Fruitery, following the line K L, Fig. 185.

ing in, the ground outside is slanted off from the walls all round (A, fig. 184), and the outside walls are constructed in cement to just above the ground.

The fruit-house is surrounded with double walls, A and B (fig. 185), having a vacant space C between them twenty inches wide. This space admirably secures the fruit from the action of the atmosphere outside.

Each of the walls is thirteen inches in thickness, made of a kind of mortar or clay formed of clayey earth, straw, and a little marl. This material is preferable to common masonry, because it is a worse conductor of heat, and is besides much cheaper. The walls should be so arranged that the floor of the vacant space between the walls, E, should be on a level with the floor of the fruit-room.

There are six openings in the walls, three in the

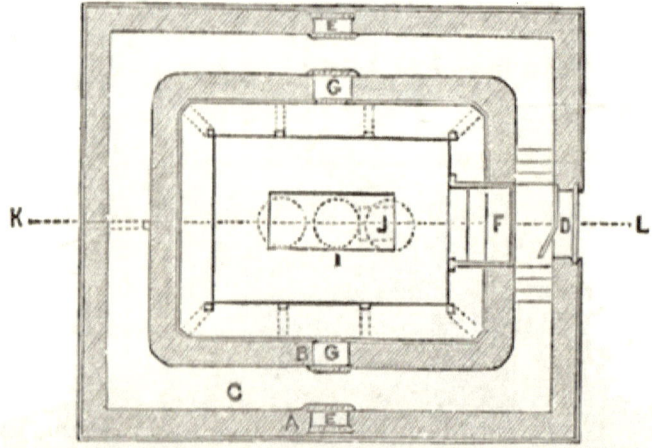

Fig. 185.—Plan of Fruitery, following the line G H, Fig. 184.

outer and three in the inner, opposite each other. The openings form:—

1st. A double door, D, fig. 185; the exterior door opens outwards, the interior one inwards, which is also made to fold like a shutter. During the extreme cold the space between the two doors must be packed with straw.

2nd. Two apertures, E, twenty inches square, placed on each side the fruit-house, and opening three feet above the ground, closed by double shutters, one opening outwards, the other inwards. The space between should be carefully packed with straw at the beginning of winter.

The inner wall has a single door, F, and two openings, G. The openings are closed by a shutter sliding up and down, and inside by a door opening outwards. As soon as the fruits are fairly deposited in the fruit-house, the exterior air should be still further excluded from the interior, by pasting strips of paper round the joints and nicks of the openings. These openings are only intended to allow of air and light being admitted into the fruit-house, to air and sweeten it *before* the fruits are deposited.

The ceiling (B, fig. 184) is formed of a bed of moss, kept in its place by laths, ceiled over above and below with mortar and hair, the whole presenting a thickness of twelve inches. This mode of structure is indispensable for preventing the exterior air from traversing the ceiling.

The ceiling is surmounted by a roof of thatch, C, twelve inches thick. A skylight in the roof will allow the space between the ceiling and the thatch being used as a garret. The skylight should close perfectly, so as to completely exclude the air.

The floor of the fruit-house may be either of wood or asphalt. The inner walls, and even the roof, will be better if lined with a deal wainscoting. All these precautions tend to the same important object, that of keeping the interior of an equable temperature, free from dampness.

All the inner walls are fitted from the top to within twenty inches of the ground with deal shelves to receive the fruit. These are placed ten inches above each other, and are twenty inches in width.

In order that all the fruit may be seen at once, the shelves are gradually raised towards the back about 45 degrees (A, fig. 186). This slant is diminished gradually as the shelves approach the floor, and at five feet above the floor they are horizontal. All the inclined shelves in front present the appearance of a

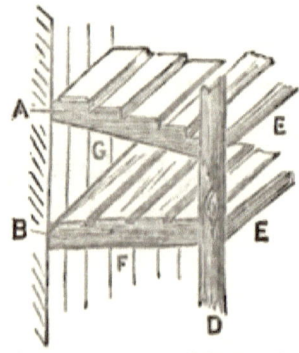

Fig. 186.—Horizontal and Inclined Shelves for the Fruitery.

rising stage or platform (A, fig. 186), each step rising about four inches, and protected in front by a ledge, E, about an inch high, to keep the fruit from falling forwards.

In order to allow the air to circulate freely, the inclined shelves are each left open behind. The same object is attained in the horizontal shelves (B), by forming them in separate leaves about four inches wide, and sufficiently apart from each other. The shelves are fixed to the wainscot by brackets, and supported in front by uprights (D) placed at five feet apart. The cross pieces (E) attached to the uprights support other cross pieces (F), both those that are horizontal and oblique, the latter cut in notches following the rise of the shelves, and upon which the shelves are placed.

In the centre of the fruit-house must be a table, I, fig. 185, ten feet long and 40 inches wide, separated from the shelves by a space of three feet. The table should be surrounded by a ledge, and the space beneath occupied by shelves placed horizontally, like the others.

It may occur that a great part of the cost of the fruit-house may be avoided. If, for instance, there is an underground cave or grotto, advantage may be taken of it to establish the fruit-house therein. In this case there will be nothing to do but to fit it up interiorly with shelves and so on, as we have just described. In every case it is indispensable that the cave or grotto be perfectly dry and well protected from the exterior temperature.

Care of the Fruit in the Fruit-house.—The success of the preservation of fruit still depends upon the care taken of it while in the fruit-house. As the fruit is brought in it is placed upon the table, which should be covered with a thin layer of dry moss. The fruit is then sorted, and each variety set apart; all the bruised and unsound fruit should be carefully put aside; the rest of the fruit should then be left upon the table for two or three days, in order to lose part of its humidity. When this time has elapsed, after covering the shelves with a thin layer of dry moss or cotton, and wiping each fruit carefully with a small piece of flannel, the fruit must be placed upon the shelves, each half an inch apart, keeping the varieties separate.

When the fruits are thus disposed of, the doors and openings must be left open during the day, at least when the weather is not too damp. Eight more days of exposure to the air are necessary, in order to allow the superabundant moisture to evaporate. Afterwards all the openings must be hermetically closed, and only opened when required to take out the fruit. No means, except currents of air, have at the present time been employed to remove dampness from the fruit-house caused by the sweating of the fruit after it has been stored. There are serious objections to the use of air currents for this object. It subjects the fruit-house to great changes of temperature, which are very injurious, and to alternate light and darkness, which hasten maturity. The plan can only be adopted in dry weather and during the absence of frost; that is, it cannot be practised throughout a great part of every winter, and the fruit-house must be left in its damp state, to the injury of the fruit.

To avoid this, we recommend the use of chloride of calcium. This salt has the property of absorbing so large a quantity of moisture (about double its own weight) that it becomes liquid after being exposed for a certain time to the influence of a moist atmosphere. We can, therefore, see that if a sufficient quantity be introduced into the fruit-house, it will absorb the dampness exhaled by the fruit. Quicklime answers nearly the same purpose.

The chloride of calcium, F, should be placed in a kind

of slanting trough, A D (fig. 187), so as to allow the chloride, as it absorbs water, to drain off into a jar, E, set underneath to receive it. This liquid should be taken care of, and when required next year placed

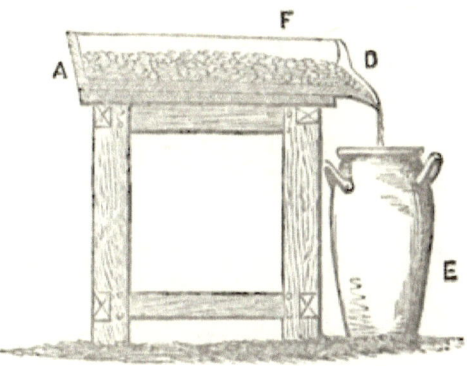

Fig. 187.—A Jar for receiving the Chloride of Calcium in the Fruitery.

upon a fire and the moisture evaporated. The residue is chloride of calcium, which may be used again as before.

The fruit-house should be visited every eighth day, to remove the fruits that are beginning to decay, to set apart those that are ripe, and to renew the chloride of calcium as may be required.

THE END.

PRINTED BY VIRTUE AND CO., CITY ROAD, LONDON.

www.ingramcontent.com/pod-product-compliance
Lightning Source LLC
Chambersburg PA
CBHW021804230426
43669CB00008B/624